SCOPE 23

The Role of Terrestrial Vegetation in the Global Carbon Cycle: Measurement by Remote Sensing

SCOPE 23

The Role of Terrestrial Vegetation in the Global Carbon Cycle: Measurement by Remote Sensing

Edited by
George M. Woodwell
The Ecosystems Center
Marine Biological Laboratory
Woods Hole, Massachusetts, USA

Published on behalf of the
Scientific Committee on Problems of the Environment (SCOPE)
of the
International Council of Scientific Unions (ICSU)
by
JOHN WILEY & SONS
Chichester · New York · Brisbane · Toronto · Singapore

Library of Congress Cataloging in Publication Data:

The Role of terrestrial vegetation in the global
  carbon cycle: Measurement by Remote Sensing

  'SCOPE 23.'
  Includes index.
  1. Carbon cycle. 2. Vegetation surveys — Remote sensing. 3. Botany — Ecology. I. Woodwell,
G.M. II. Scientific Committee on Problems of the Environment.
QH344.R64   1984   574.5′22   83-10333

ISBN 0 471 90262 4

British Library Cataloguing in Publication Data:

The Role of terrestrial vegetation in the global carbon
  cycle: measurement by remote sensing.
  1. Carbon cycle (Biogeochemistry — Contresses
  I. Woodwell, George M.
  574.5′2   QH344

ISBN 0 471 90262 4

Filmset by Mid-County Press, London SW15
and printed by the Pitman Press Ltd, Bath, Avon.

International Council of Scientific Unions (ICSU)
Scientific Committee on Problems of the Environment (SCOPE)

SCOPE is one of a number of committees established by a non-governmental group of scientific organizations, the International Council of Scientific Unions (ICSU). The membership of ICSU includes representatives from 68 National Academies of Science, 18 International Unions and 12 other bodies called Scientific Associates. To cover multidisciplinary activities which include the interests of several unions, ICSU has established 10 scientific committees, of which SCOPE, founded in 1969, is one. Currently, representatives of 34 member countries and 15 Unions and Scientific Committees participate in the work of SCOPE, which directs particular attention to the needs of developing countries.

The mandate of SCOPE is to assemble, review, and assess the information available on man-made environmental changes and the effects of these changes on man; to assess and evaluate the methodologies of measurement of environmental parameters; to provide an intelligence service on current research; and by the recruitment of the best available scientific information and constructive thinking to establish itself as a corpus of informed advice for the benefit of centres of fundamental research and of organizations and agencies operationally engaged in studies of the environment.

SCOPE is governed by a General Assembly, which meets every three years. Between such meetings its activities are directed by the Executive Committee.

R.E. Munn
Editor-in-Chief
SCOPE Publications

Executive Secretary: V. Plocq

Secretariat: 51 Bld de Montmorency
75016 PARIS

SCOPE 21: The Major Biogeochemical Cycles and Their Interactions, 1983, 554pp

SCOPE 22: Effects of Pollutants at the Ecosystem Level, 1984

SCOPE 23: The Role of Terrestiral Vegetation in the Global Carbon Cycle: Measurement by Remote Sensing, 1984

Funds to meet SCOPE expenses are provided by contributions from SCOPE National Committees, an annual subvention from ICSU (and through ICSU, from UNESCO), an annual subvention from the French Ministere de l'Environnement et du Cadre de Vie, contracts with UN Bodies, particularly UNEP, and grants from Foundations and industrial enterprises.

Contents

ix

SECTION IV: REMOTE SENSING

Contents   xiii

SECTION V: CONCLUSION

Preface

More than 99 per cent of the volume of the atmosphere is two gases, nitrogen and oxygen. The remainder is a group of trace gases among which carbon dioxide (CO_2) is one of the most abundant. The amount of carbon dioxide is important because, in contrast to nitrogen and oxygen, carbon dioxide absorbs radiant heat (infra-red energy). A change in the amount of CO_2 in the atmosphere has the potential for changing the amount of radiant energy retained in the atmosphere, thereby changing climates globally.

Over the past century or more the CO_2 content of the atmosphere has been increasing. The rise since 1900 is thought to have been 10–20 per cent, although it is not known with precision. In 1980 the total CO_2 in the atmosphere was about 336 p.p.m. by volume. If the present rate of increase continues we can expect about 600 p.p.m. in the atmosphere by the middle of the next century, possibly sooner. Such a rise is expected to cause a warming of the earth that may average as much as 1.5–4.0 °C. The warming will not be uniform. There will be little change in the tropics and a 6–8 °C rise in the polar regions. Such a change in temperature will shift climatic zones, disrupt agriculture, and, over a century or so, melt sufficient polar ice to raise sea-level one to five metres. Although absolute proof that these changes in climate will occur is lacking, and may not be available soon, there is powerful evidence for an influence of CO_2 on temperature of the magnitude indicated. Other factors such as the output of the sun and the amounts of particulate matter in the atmosphere may nonetheless dominate trends of climate.

The cause of the increase in CO_2 in the atmosphere is both the combustion of fossil fuels and the steady destruction of forests. The latter process has been under way for centuries but has recently been accelerated; the former is a product of the last two centuries of the industrial revolution.

The changes in climate that the increase in CO_2 threatens is ample reason for intensive efforts at improving predictions of future CO_2 concentrations. The accuracy of predictions is dependent in large part on knowledge of the sources of CO_2 released into the atmosphere. The amount released from combustion of fossil fuels is known with considerable precision. The amount from the biota and soils of the earth is uncertain. The growth of the human population to more than six billion by the year 2000 seems to assure a demand both for wood and for the continued expansion of agriculture onto forested lands. The rate of these changes is the primary datum for appraising the importance of the biota in affecting the CO_2 content of the atmosphere. The question addressed in this book is how to determine that rate for the world as a whole.

The book is the product of a SCOPE conference* arranged by The Ecosystems Center and held in Woods Hole, Massachusetts in May, 1979. Further analysis, stimulated in part by the conference and in part by research suggested by the conference, led to substantial advances during the period of preparation of the report and this book. The Report of the conference has been published as DOE Publication _CONF-7905176_ (1980). This book includes a series of review papers that carry the progress in research through 1982 into the early months of 1983.

Editor's note

Throughout this volume the term 'billion' is used to mean one thousand million.

* The Conference was under the auspices of the Scientific Committee on Problems of the Environment of the International Council of Scientific Unions. Financing was in part by a grant from the Department of Energy, in part by a grant from the Exxon Corporation, and in part by funds from the Ecosystems Center.

List of Contributors

F. C. Billingsley   Jet Propulsion Laboratory, California Institute of Technology, Pasadena, California 91103, USA

P. Buringh   Marterlaan 20, 6705 BA Wageningen. The Netherlands

J. D. Erickson   NASA Johnson Space Center, Houston, Texas 77058, USA

J. E. Hobbie   The Ecosystem Centre, Marine Biological Laboratory, Woods Hole, Massachusetts 02543, USA

R. M. Hoffer   Department of Forestry and Natural Resources, Purdue University, West Lafayette, Indiana 47907, USA

R. A. Houghton   The Ecosystems Center, Marine Biological Laboratory, Woods Hole, Massachusetts 02543, USA

J. M. Melillo   The Ecosystems Center, Marine Biological Laboratory, Woods Hole, Massachusetts 02543, USA

B. Moore III   Complex Systems Center, University of New Hampshire, Durham, New Hampshire 03824, USA

D. Mueller-Dombois   Department of Botany, University of Hawaii at Manoa, Honolulu, Hawaii 96882, USA

A. B. Park   606 Shore Acres Road, Arnold, Maryland 21012, USA

B. J. Peterson   The Ecosystems Center, Marine Biological Laboratory, Woods Hole, Massachusetts 02543, USA

W. H. Schlesinger   Department of Botany, Duke University, Durham, North Carolina 27706, USA

G. R. Shaver   The Ecosystems Center, Marine Biological Laboratory, Woods Hole, Massachusetts 02543, USA

G. M. Woodwell   The Ecosystems Center, Marine Biological Laboratory, Woods Hole, Massachusetts 02543, USA

SECTION I

Introduction

The Role of Terrestrial Vegetation in the Global Carbon Cycle:
Measurement by Remote Sensing
Edited by G. M. Woodwell
© 1984 SCOPE. Published by John Wiley & Sons Ltd

CHAPTER 1
The Carbon Dioxide Problem

G. M. WOODWELL
*The Ecosystems Center, Marine Biological Laboratory, Woods Hole,
Massachusetts, USA*

1.1 THE ROLE OF THE TERRESTRIAL BIOTA

The amount of carbon dioxide (CO_2) in the atmosphere is thought to have increased by 40–80 parts per million (p.p.m.) since 1860, although the data prior to 1958 are uncertain. In 1958 Keeling and colleagues started systematic measurements of CO_2 at an elevation of 3400 m on Mauna Loa, a volcanic mountain on the island of Hawaii (Pales and Keeling, 1965; Keeling *et al.*, 1976b). Mauna Loa was chosen because its remoteness from urban centres made possible the sampling of well-mixed air of the troposphere. The record from Mauna Loa and a parallel record also started in 1958 at the South Pole (Keeling *et al.*, 1976a), together with data from elsewhere (Bolin and Bischof, 1970; Woodwell *et al.*, 1973; Machta *et al.*, 1977; Pearman, 1980), are conclusive evidence that the CO_2 content of the atmosphere is rising annually. In 1980 the atmosphere contained about 336 p.p.m. and the amount was increasing at 1.0 to 1.5 p.p.m. annually.

The measurements also show that the concentration of CO_2 varies in a regular pattern. All continuous records such as those from Mauna Loa show a seasonal oscillation with a peak in late winter and a minimum in late summer. The variation is thought to be caused by the metabolic activity of temperate zone forests. During the summer in both hemispheres there is net storage of carbon because photosynthesis exceeds respiration. During the autumn, winter and spring there is a net release of carbon because total respiration exceeds gross photosynthesis. The pattern produces oscillations in the CO_2 content of the atmosphere that vary in amplitude with latitude and elevation (Bolin and Keeling, 1963; Bolin and Bischof, 1970; Machta *et al.*, 1977; Pearman, 1980). At Mauna Loa the amplitude is about five p.p.m. The pattern is reversed in the southern hemisphere to follow the southern seasons.

The year-by-year increase in CO_2 is the product of a series of interactions between the atmosphere, the oceans, the terrestrial biota and human activities. There are two important sources of CO_2: the combustion of fossil fuels and

3

the decay (or combustion) of biotic residues on land. The most important biotic source is the destruction of forests. Some of the carbon from these two sources accumulates in the atmosphere; some is transferred to the oceans. In the long-term of centuries to millennia, equilibrium would be expected between the rate of release, the atmospheric concentration, and the oceanic concentration. Currently, however, the rate of transfer of carbon into the oceans is less than the rate of release of CO_2 into the atmosphere. The result is the annual increase we observe in the atmosphere of 1.0–1.5 p.p.m./yr.

The relationships between these factors have been summarized by Woodwell *et al.* (1983b) for 1980. The annual increase in CO_2 in the atmosphere is the difference between the total release and the transfer to the oceans, expressed as:

Atmospheric Increase = Fossil Release − Oceanic Absorption ± Biotic Effect

In this equation the atmospheric increase is known to be 1.0–1.5 p.p.m. or $2.0–3.0 \times 10^{15}$ g C/yr.* The release from fossil fuels, 5.2×10^{15} g in 1980, is thought to be known within ± 15 per cent or less (Rotty, 1982). A review of the biotic contribution (Houghton *et al.*, 1983) suggests a further net release to the atmosphere of $2–5 \times 10^{15}$ g C annually from deforestation. The equation is obviously not balanced. In an attempt to establish the degree of uncertainty associated with evaluation of the equation Woodwell *et al.* (1983b) showed that the range of imbalance in 1980 was from about 1.5×10^{15} g C to about 7×10^{15} g. The greatest uncertainty was associated with the role of the biota and the magnitude of the absorption by the oceans.

The total amount of carbon held in the biota and in the organic matter of soils has been estimated at between 2000 and 4000×10^{15} g (Table 1.1). A small change in the amount of carbon stored on land might change the atmospheric concentration appreciably. For instance a change of 0.1 per cent in a terrestrial pool of 2000×10^{15} g C would be approximately equivalent to a one p.p.m. change in the total atmospheric burden. The extent to which the biota and the humus of the earth as a whole are changing is obviously important in trying to predict the future CO_2 content of the atmosphere.

Most of the carbon stored in the earth's biota and soils is associated with forests (Table 1.1) (Whittaker and Likens, 1973; Rodin *et al.*, 1975; Olson *et al.*, 1978; Ajtay *et al.*, 1979). A change in the area of land covered by forests world-wide or a change in the stature of forests due to a shift in the ratio of the gross photosynthesis to total respiration of terrestrial ecosystems (Woodwell and Whittaker, 1968) might be expected to produce a net exchange of carbon with the atmosphere in the same range as the $5–6 \times 10^{15}$ g released in 1980 from combustion of fossil fuels.

Predictions of the future CO_2 content of the atmosphere have been based

---

* 1 billion metric tons = 10^{15} g.

Table 1.1   Estimates of the terrestrial carbon pools according to various authors

Vegetation	Soils $(10^{15}$ g C)	Reference
827		Whittaker and Likens, 1973
557		Olson *et al.*, 1978
559.8	2070	Ajtay *et al.*, 1979
	1515	Schlesinger, 1983 (this volume)
	3000	Bohn, 1978
	1477	Buringh, 1983 (this volume)

	Ranges	
Vegetation	Soils	Total
557–827	1500–3000	2000–3800

primarily on projections of data from Mauna Loa and elsewhere. Such predictions are tentative because knowledge of the factors that determine the amount of CO_2 in the atmosphere remains inadequate to explain details of the global carbon cycle. Models used for such projections since 1970 (SCEP, 1970) have usually incorporated the assumption that the oceans have a limited capacity for absorbing atmospheric CO_2, at least over a period of a few decades, and that the biota is also absorbing atmospheric carbon (Machta, 1973; Bacastow and Keeling, 1973; Oeschger *et al.*, 1975; Broecker *et al.*, 1979). More recent analyses suggest that the biota is not currently a net accumulator of atmospheric CO_2 but is releasing stored carbon into the atmosphere (Adams *et al.*, 1977; Bolin, 1977; Woodwell *et al.*, 1978; Wong, 1978; Hampicke, 1979; Moore *et al.*, 1981; Houghton *et al.*, 1983; Woodwell *et al.*, 1983b). Interpretations of the world carbon cycle based on the earlier models appear to be in error. There is a need for greater accuracy in knowledge of the factors in the equation, especially the role of the biota, if we are to make accurate predictions and to consider action to avoid global climatic change. This book is a step toward development of that information.

1.2   FACTORS AFFECTING THE STORAGE AND RELEASE OF CARBON

The amount of carbon retained in the biota and terrestrial humus world-wide is affected by many factors, the most important of which are changes in the area of forests due to human activities and any regional shift in the ratio of gross photosynthesis to total respiration.

Changes in the area of forests occur as a result of several deliberate or inadvertent human activities:

(a) *Harvest of forests*
  The harvesting of forests by clear-cutting is the most easily observed but not the most common technique in managing forests. Partial cuttings are far more common, and measurement of the remaining carbon contained in the residual forest and soil is difficult. Measurement is further complicated by the fact that forests normally recover over time through succession.

(b) *Expansion of agricultural land into forested land*
  This transition results not only in the loss of the large pool of carbon held in the biota, but also in the loss over a few years of the carbon held in forest soils.

(c) *Expansion of grazing lands*
  As population expands forests come under increasing pressure not only for agricultural lands but also for grazing. The transformation is largely unmeasured, but it is also important in determining the amount of carbon released to the atmosphere.

(d) *Biotic impoverishment*
  Intensive farming, repeated burning of vegetation, grazing and other agricultural practices often result in changes in the vegetation and soils that preclude the re-establishment of forests. These changes constitute long-term impoverishment and result in a net transfer of carbon to the atmosphere.

(e) *Abandonment of land*
  The abandonment of agricultural lands in forested zones of the earth usually leads to the re-establishment of forests through succession. Such transitions have occurred over large areas in eastern North America as agriculture has become concentrated on richer lands further west. The increase in forested land in the East has resulted in the net storage of carbon in that area (Delcourt and Harris, 1980).

(f) *Recovery from harvest or other disturbance*
  Partial harvests of forests, fires and other disturbances start successional changes that result in the accumulation of carbon in the biota and in soils. Measurement of these changes is possible but not simple.

The ratio of gross photosynthesis to total respiration can also be affected by several factors. The topic has been reviewed recently by Woodwell (1983). The greatest sensitivity of this ratio is probably in response to temperature. Respiration is especially responsive: an increase of 10 °C commonly increases rates of respiration by two to three times. Effects of temperature on photosynthesis are usually indirect through effects on respiration; direct effects are small.*A short-term warming, unaccompanied by other climatic changes, can be expected to shift the ratio of CO_2 exchange in terrestrial ecosystems in favor of respiration, at least in the period of a few years, and result in a net release of carbon from the biota to the atmosphere. A cooling would work in the opposite direction. In the longer term of decades to centuries the new climatic regimes will result in migration of species and development of new natural communities.

Much attention has been given to the potential of an increase in CO_2 in the atmosphere for stimulating gross photosynthesis globally and thereby causing an increase in the storage of carbon in terrestrial systems. The possibility was formalized in a well-known model of the global carbon cycle by Bacastow and Keeling (1973) as follows:

$$F_{ab} = F_{bo}\left[1 + \beta \ln\left(\frac{N_a}{N_{ao}}\right)\right]\left(\frac{N_b}{N_{bo}}\right)$$

where F_{ab} is the flux of carbon from the atmosphere to the biota; F_{bo}, the preindustrial value of F_{ab}; N_a and N_b, the carbon in the atmosphere and the biota; and N_{ao} and N_{bo}, their preindustrial values was called the biota growth factor; it defines the degree of CO_2 fertilization.

There is no direct evidence from observation of natural communities that the increase in CO_2 has resulted in an increase in storage of carbon globally. The factors that affect the accumulation of carbon in natural communities are sufficiently numerous and complex to lead to considerable doubt as to whether such stimulation can occur as a general rule (Kramer, 1981; Woodwell, 1983; Woodwell *et al.*, 1983b). Small changes in temperature probably have a greater effect than a change in CO_2 concentration. Although the β factor, loosely defined, has been used in models, its use has been as an adjustment in a model, not as a description of a process. It should be replaced by consideration of both the effects of temperature where appropriate and the effects of the transformation or harvesting of forests.

These analyses have focused attention for the moment on measurement of changes in the vegetation of the earth, especially changes in the area of forests, as the most important step toward improvement in knowledge of the global carbon cycle. Other factors affect the cycle and influence the CO_2 content of the atmosphere. But the estimates of rates of deforestation vary greatly. The estimates for the globe as a whole appear in Table 1.2. They range for the mid-seventies from a net loss of about 3.3×10^6 ha (FAO, 1977) to 20.1×10^6 ha (Barney, 1980). If a clarification of the role of the biota and the oceans is to occur and be sufficient for improved prediction and, presumably, control of atmospheric CO_2, the first step is development of a simple system for measuring and tabulating changes in the vegetation of the earth. Basic elements

Table 1.2   Range of estimates of clearing rates of tropical and subtropical closed forests using various sources (adapted from Woodwell *et al.*, 1983b)

Clearing rate estimate per year ($\times 10^6$ ha)	Source
3.3	FAO, *Production Yearbooks* (1977)
4.2	Population-based estimate (Houghton *et al.*, 1983)
7.5	FAO—Forestry Section (1982)
7.8	Brown and Lugo (1980)
13.1	Myers (1980)
20.1	Global 2000 (Barney, 1980)

of such a system are available now. They include both methods for using remote sensing as the primary source of new data and a model designed to keep a record of changes in the vegetation regionally and globally.

1.3   A PRACTICAL APPROACH TO MEASUREMENT OF CHANGES IN THE AREA OF VEGETATION

1.3.1   Satellite Imagery

Remote sensing offers three potential methods for measuring changes in terrestrial vegetation: the single image inventory, sequential-image inventories, and paired-image detection of change. These methods have been examined in detail in a report to the US Department of Energy by Woodwell *et al.* (1983a). They are summarized below.

1.3.1.1   Single Image Inventory

Ecologists have long recognized an ability to 'read the landscape'. A knowledge of the successional relationships among plant communities, of the history of use of the land in any area, and of the factors that are important in affecting the vegetation at any moment enable ecologists to interpret changes and rates of change in a landscape with considerable accuracy. These interpretations can be formalized, made quantitative, and used to describe the net flux of carbon between the landscape and the atmosphere. The question is whether that ability can be developed with remotely-sensed imagery as the basis for interpreting the landscape.

The general answer is certainly affirmative, but the effort may be very large and require more experience with the vegetation than is practical and more detail in the imagery than is possible. Nonetheless, a single image can be classified and mapped, the various types of vegetation and other land classes can be defined. If two per cent of the area has been transformed to non-forest and moves through succession into a different class in two years, the rate of production of 'non-forest' is about one per cent annually.

The difficulties are in the classification and interpretation; a great deal of knowledge is required for accuracy. A considerably larger number of classes than the four to ten that are practical with LANDSAT imagery would be required for the types of classification required by Mueller-Dombois (this volume, pp. 21–83) and for the purpose outlined here. An approach based on classification of a single image from LANDSAT appears impractical.

1.3.1.2 Sequential Image Inventory

Superficial analyses of the challenge of measuring a change in the amount of carbon in the biota and soils globally usually suggest a need for a comparison of total inventories taken at different times. Such inventories are possible, of course, but require the highly accurate classification discussed above. If errors are made in the classifications, the errors appear in the estimate of change and contribute to its uncertainty. The difficulties inherent in the single image inventory are intrinsic in the sequential image inventory and limit its potential to very special circumstances that probably cannot involve LANDSAT imagery.

1.3.1.3 Change Detection with Paired Images

If the objective can be limited to measurement of the extent to which forests have been changed to non-forest and non-forest has reverted to forest over a period of a few years, the analysis by satellite imagery can be simplified. One possibility is to compare images of the same place taken at different times. If the images can be superimposed accurately, it is possible to identify those areas where the change has occurred and to eliminate from further consideration all other areas. A third image can be prepared that contains only the areas that have been changed. The problem of classification is reduced to that of identifying the character of the change. And the only measurements made are those of areas that have changed. Errors of classification and measurement are minimized. The technique has been applied as a test in the forests of Maine and is discussed in Woodwell *et al.* (this volume, pp. 221–240).

1.3.2 Tabulating Changes: a Simple Model

The first step in using data on changes in vegetation is the development of a method of recording the changes as they occur. The most important changes are the transformation of forest to non-forest, either through clear-cutting or transformation to agriculture. Once such a change has occurred, if the land does not remain under cultivation, plant succession starts. Because succession follows a predictable course we need know only the fact that succession has, or has not, been started; we can calculate its rate of progress from previous experience; no further measurement is required. The sampling programme is reduced to the detection of change as opposed to the extensive classification required for the techniques based on inventory, outlined by Woodwell *et al.*, this volume.

1.4   THE CLASSIFICATION OF VEGETATION

1.4.1   History, Basic Principles and Trends

Dieter Mueller-Dombois of the University of Hawaii offers a comprehensive summary of the many systems that have been developed for classifying vegetation. Mueller-Dombois shows how the older systems of classification have now been developed to a high degree of complexity to provide a variety of methods for classification and mapping plant communities. These advances have been fundamental to explaining causes of patterns in vegetation.

The earlier work makes possible the mapping of phytomass (plant mass), primary production and losses of vegetation in forests. Thus far, little work has been done in making these determinations in tropical areas. Future efforts will be complicated by the diversity of plant species within the tropics and the absence of dominance. Both factors complicate classification and the preparation of maps.

Although remote sensing by satellites has greatly advanced our ability to obtain data on vegetation, Mueller-Dombois reports that the intermediate scale imagery available now can be used primarily to differentiate only between broad types of vegetation. In certain instances this imagery has been enlarged successfully to show greater detail. Nonetheless, most intermediate scale imagery from satellites presents only the basic structural elements of vegetation, such as shape, texture, shadows, and albedo. Mueller-Dombois suggests that the most useful approach to vegetation inventory is to use satellite imagery in combination with large-scale maps, such as those obtained through aerial photography, with other types of information.

1.5   SOILS

1.5.1   An Emphasis on Organic Matter in Forests

Soils develop with the vegetation. The organic content of soils, although not measurable directly by remote sensing, can be inferred from knowledge of the vegetation. Two papers in this volume treat the changes in the amount of carbon retained in soils of the earth as a whole. Paul Buringh of the Agricultural University of the Netherlands at Wageningen has provided a systematic analysis of the decline in organic carbon in soils of the world. William Schlesinger of Duke University, Durham, North Carolina has made a parallel inventory of organic carbon in soils. Most of the decline in organic carbon in soils occurs following the transition from forest to agriculture or other non-forest uses. The papers arrive at substantially different conclusions concerning the magnitude of the loss of carbon following disturbances and thereby emphasize the difficulty in making such estimates. The pattern of change, however, is consistent between these papers and with other analyses.

Buringh uses the eleven major soil types of the US Department of Agriculture to show that a 'realistic estimate' of the amount of carbon lost from soils each year is 4.6×10^{15} g. This loss represents 0.3 per cent of the estimate of currently existing soil carbon accepted by Buringh, 1477×10^{15} g. The loss is unevenly distributed throughout the world with the largest changes in the tropics. Each year, for example, he suggests that 24×10^6 ha of forests are cut in the Far East, Latin America, Africa and Oceania to support an estimated 250 million people who depend on shifting cultivation. In addition, forest fires are estimated globally to destroy 5×10^6 ha of forests with their soils each year. The figures exceed the estimates of Table 1.1, but their source is unclear.

Buringh estimates that the carbon content of the earth's soils in prehistoric times was 2014×10^{15} g. Of that, 537×10^{15} g, or 27 per cent of the prehistoric total, have how been lost. He argues that 'it would be much better to intensify agricultural production on presently cultivated land than to convert more forestland to food production'.

Schlesinger has revised his 1977 estimate of the amount of carbon in the soil, using 35 new values for soil carbon in addition to the 82 values used for the 1977 estimate and also using a figure of 7.9 kg C/m^2 as the value of carbon in cultivated soils rather than the figure of 12.7 kg C/m^2. The latter figure was based on the assumption that most cultivated land contains the same amount of carbon as temperate zone grassland. On the basis of work by Revelle and Munk (1977), Schlesinger now assumes that cultivated soils lose 40 per cent of their carbon during cultivation.

Despite these changes in assumptions, Schlesinger's new estimate of the total carbon in the world's soils of 1515×10^{15} g of carbon is only 4 per cent higher than his 1977 estimate. He estimates that the current release of carbon to the atmosphere from the world's soils is 0.8×10^{15} g per year, and that the total cumulative transfer of carbon from soils to the atmosphere since prehistoric times may have reached 40×10^{15} g.

It is clear from these discussions that there is a roughly fivefold variation among current estimates of releases from soils. The remaining papers in this book treat details of remote sensing techniques and experience in applying them to problems closely related to those set forth above.

1.6   REMOTE SENSING

1.6.1   How Does it Work?

R. M. Hoffer of Purdue University provides an analysis of the three types of remote sensing data used to classify vegetation: aerial photographs, multispectral scanners (MSS) that record spectral characteristics, and radar imagery.

Each of these methods has particular advantages and disadvantages in interpreting plant cover and our purposes may require use of two or all three of these methods in combination. Large-scale and intermediate-scale photographs obtained with standard colour film, for example, are quite suitable for identifying individual species of trees, and similar scale photographs obtained with colour infra-red film are suitable for identifying grasses and herbs. Such photographs, however, cover small areas and must be interpreted using manual methods. Photographs and images from satellites, on the other hand, provide views of much larger areas, although they do not provide as much detail.

The multispectral scanners (MSS) are commonly used on satellites. The data are virtually instantaneous records of the energy reflected or emitted in various portions of the electromagnetic spectrum by objects on earth and are ideally suited for processing by computer. Although MSS data often cannot be used to distinguish between different types of vegetation, such data can now be used to distinguish large areas of forestland, rangeland and agricultural land. MSS data also are sufficient to distinguish between deciduous and coniferous forests and large areas in a uniform crop.

Radar has several advantages because it is an active rather than a passive system. One advantage is that radar detects objects on the ground through cloud cover and at night. In addition, side-looking airborne radar (SLAR) systems are particularly appropriate for distinguishing between the physiognomy of different classes of vegetation. Such changes in vegetation as the recent clearing of forests are also distinct on radar imagery.

F. C. Billingsley of the National Aeronautics and Space Administration (NASA) provides a more detailed discussion of the value of the multispectral scanner with computer enhancement in providing detailed imagery from satellites. He outlines the problems that confront the designer of multispectral scanning devices. Billingsley emphasizes that clouds and other problems with the scanners ensure that microwave sensing by radar will be necessary to collect the data needed to determine changes in the earth's vegetation in certain places. Radar, he says, will continue to be particularly valuable for measuring stand density, delineating forest lands from shrublands and pasture, differentiating between tree species, making inventories of tropical forests and measuring tree heights. Accuracy in identifying vegetative cover may be enhanced by careful selection of spectral bands. In the new scanners narrower bands will be available as the technology of linear arrays develops. These scanners will improve resolution considerably, but probably not enough to satisfy the needs for classification as outlined by Mueller-Dombois.

The ultimate test of remotely sensed data is in their usefulness in interpreting the surface of the earth. A. B. Park of the General Electric Company emphasizes the complexities intrinsic in making inferences from imagery. There is no substitute for experience on the ground in Park's eyes. He observed that all remotely sensed images are inevitably subject to some degree of error caused by several factors such as: constant changes in the intervening atmosphere; variation in the calibration and in the resolving power of the instrument; the influence of time of day and season on the material measured; and the inability to observe the target material in its pure state in most situations. In short, accuracy and reproducibility in the remote sensing of materials on earth are elusive, although the accuracy can be high for such uncomplicated features as bare soil and water.

Correlating remotely sensed images with materials on earth, Park observes, requires substantial information that must be obtained through detailed study of the land. The richer the array of collateral information, the more effective the classification. Park concludes that the coupling of remotely sensed data with ground observations is a task only for scientists with training and experience.

A detailed examination of an experiment that applied MSS data in combination with meteorological, climatological and conventional data, to predict the growth of wheat in the United States, the USSR and other countries is found in the paper by Jon D. Erickson of the National Aeronautics and Space Administration (NASA).

The basic objective of the Large Area Crop Inventory Experiment (LACIE), was to develop and test a method of estimating world-wide production of wheat. The technique was to use multispectral scanners in satellites to obtain information on selected segments of land for the purpose of estimating the area of land planted in wheat, and combining these estimates of area with estimates of yield based on models that relied on weather data.

Carried out during the global crop years 1974–75 (Phase I), 1975–76 (Phase II) and 1976–77 (Phase III), LACIE's most notable accomplishments were accurate estimates of US winter wheat production for the three crop years and, in 1977, an accurate estimate of a serious shortfall in the Soviet spring wheat crop long before the USSR released similar information. The final LACIE estimate differed from the final production figure released by the USSR by only one per cent, a remarkable achievement given the extreme variability in the production of most large-scale crops.

LACIE was not always so successful. During Phase II the experiment encountered considerable difficulty in distinguishing spring wheat from other small spring grains in the United States and Canada. The difficulty led to underestimates of production for the spring wheat area of both the United States and Canada.

Nonetheless, the LACIE experiment was a huge step forward in the development of techniques for estimating overall production on the basis of area and yield estimates, for estimating planted areas without the use of confirming ground data, and for estimating crop yields per acre. As a result, efforts are now under way to extend the prediction through the LACIE experiment to other important crops.

The evidence from these several papers treating remote sensing re-emphasizes the need for clarity and simplicity in objectives. The emphasis on an approach based on measurements of changes as opposed to classification and successive inventories seems most appropriate.

1.6.2   Practical Experience: a Successful Test

The final chapter by a group at The Ecosystems Center in Woods Hole offers an analysis of one potential approach to the use of LANDSAT imagery in the measurement of changes in the vegetation of the earth. The system is based on paired LANDSAT images from different times, superimposed to detect the areas changed.

The technique requires use of a model designed to maintain the record of when and where the change occurred, the stage of the recovery through plant succession, and the net exchange of carbon with the atmosphere. Details of a model developed for this purpose have been presented by Moore *et al.* (1981), Houghton *et al.* (1983) and Woodwell *et al.* (1983b), who have shown its application on the basis of published data. The model was designed, however, to accept data based on satellite imagery or other sources. The purpose was to determine the net biotic exchange of CO_2 with the atmosphere over the past century or two. To this end, 10 different geographic regions and 12 types of vegetation were accommodated. The model could be modified for different initial amounts of carbon, different periods of plant succession, different degrees of recovery following harvest, and for harvests of varying intensities. The most important information was (a) a measurement of the rate of transformation of forests to non-forests and (b) knowledge of the fate of the tract, that is, whether it was allowed to return to forest, was tilled or grazed, became impoverished, or was used in some other way (Houghton *et al.*, 1983). Once these two facts were known, the model could be used to provide a detailed record of carbon storage for that site, for each region, and for the earth as a whole.

Experience suggests that this system will work. It will probably require much ancillary information from the areas of interest and may require radar imagery as well. The potential value of this approach extends well beyond the immediate needs of clarifying aspects of the global carbon cycle to the challenges inherent in managing resources globally for a human population expected to surge well beyond six billion by the early years of the next century.

1.7 REFERENCES

Adams, J. A. S., Mantovani, M. S. M., and Lundell, L. L. (1977) Wood versus fossil fuel as a source of excess carbon dioxide in the atmosphere: a preliminary report. *Science*, **196**, 54–56.

Ajtay, G. L., Ketner, P., and Duvigneaud, P. (1979) Terrestrial primary production and phytomass. In: Bolin, B., Degens, E. T., Kemps, S., and Ketner, P. (eds), *The Global Carbon Cucle, SCOPE 13*, 129–182. John Wiley and Sons, New York.

Bacastow, R., and Keeling, C. D. (1973) Atmospheric carbon dioxide and radiocarbon in the natural carbon cycle. II. Changes from AD 1700 to 2070 as deduced from a geochemical model. In: Woodwell, G. M., and Pecan, E. V. (eds), *Carbon and the Biosphere*. CONF-720510 AEC Symposium Series 30, 86–135. Technical Information Center, Oak Ridge, Tennessee.

Barney, G. O. (ed) (1980) *The Global 2000 Report to the President of the U.S.—Entering the 21st Century*. Pergamon Press, New York.

Bohn, H. L. (1978) Organic soil carbon and CO_2. *Tellus*, **30**, 472–475.

Bolin, B. (1977) Changes of land biota and their importance for the carbon cycle. *Science*, **196**, 613–615.

Bolin, B., and Bischof, W. (1970) Variations of the carbon dioxide content of the atmosphere in the northern hemisphere. *Tellus*, **22**, 431–442.

Bolin, B., and Keeling, C. D. (1963) Large-scale atmospheric mixing as deduced from the seasonal and meridional variations of carbon dioxide. *J. Geophys. Res.*, **68**, 3899–3920.

Broecker, W. S., Takashashi, T., Simpson, H. J., and Peng, T. H. (1979) Fate of fossil fuel carbon dioxide and the global carbon budget. *Science*, **206**, 409–418.

Brown, S., and Lugo, A. E. (1980) The role of terrestrial biota on the global CO_2 cycle. Proceedings of symposium: a review of the carbon dioxide problem. *Am. Chem. Soc., Div. Pet. Chem.*, **26**, 1019–1025.

Delcourt, H. R., and Harris, W. F. (1980) Carbon budget of the southeastern U.S. biota: analysis of historical change in trend from source to sink. *Science*, **210**, 321–322.

FAO (1977) *Production Yearbook*. FAO, Rome, Italy.

FAO Forestry Staff (1982) Tropical Forest Resources. FAO, Rome, Italy.

Hampicke, U. (1979) Net transfer of carbon between the land biota and the atmosphere, induced by man. In: Bolin, B., Degens, E. J., Kempe, S., and Ketner, P. (eds), *The Global Carbon Cycle, SCOPE 13*, 219–236. John Wiley and Sons, New York.

Houghton, R. A., Hobbie, J. E., Melillo, J. M., Moore, B., Peterson, B. J., Shaver, G. R., and Woodwell, G. M. (1983) Changes in the carbon content of terrestrial biota and soils between 1860 and 1980: a net release of CO_2 to the atmosphere. *Ecol. Monogr.* (In press).

Keeling, C. D., Adams, J. A. Jr., Ekdahl, C. A. Jr., and Guenther, P. R. (1976a) Atmospheric carbon dioxide variations at the South Pole. *Tellus*, **28**, 552–564.

Keeling, C. D., Bacastow, R. B., Bainbridge, A. E., Ekdahl, C. A. Jr., Guenther, P. R., and Waterman, L. S. (1976b) Atmospheric carbon dioxide variations at Mauna Loa Observatory, Hawaii. *Tellus*, **28**, 538–551.

Kramer, P. J. (1981) Carbon dioxide concentration, photosynthesis and dry matter production. *BioScience*, **31**, 29–33.

Machta, L. (1973) Prediction of CO_2 in the atmosphere. In: Woodwell, G. M. and Pecan, E. V. (eds), *Carbon and the Biosphere*. AEC Symposium Series 20, 21–31. Technical Information Center, Springfield, Virginia.

Machta, L., Hanson, K., and Keeling, C. D. (1977) Atmospheric carbon dioxide and some interpretations. In: Andersen, N. R. and Malahoff, A. (eds), *The Fate of Fossil Fuel CO_2 in the Oceans*, 131–144. Plenum Press, New York.

Moore, B., Boone, R. D., Hobbie, J. E., Houghton, R. A., Melillo, J. M., Peterson, B. J., Shaver, G. R., Vorosmarty, C. J., and Woodwell, G. M. (1981) A simple model for analysis of the role of terrestrial ecosystems in the global carbon budget. In Bolin, B. (ed.), *Carbon Cycle Modelling, SCOPE 16*, 365–385. John Wiley and Sons, New York.

Myers, N. (1980) Report of survey of conversion rates in tropical moist forests. National Research Council, Washington, DC. 205 p.

Oeschger, H., Siegenthaler, U., Schotterer, U., and Gugelmann, A. (1975) A box diffusion model to study the carbon dioxide exchange in nature. *Tellus*, **27**(2), 168–192.

Olson, J. S., Pfuderer, H. A., and Chan, Y. H. (1978) Changes in the global carbon cycle and the biosphere. ORNL/ETS-109, Oak Ridge National Laboratory, Oak Ridge, Tennessee. 169 p.

Pales, J. C., and Keeling, C. D. (1965) The concentration of atmospheric carbon dioxide in Hawaii. *J. Geophys. Res.*, **70**, 6053–6076.

Pearman, G. I. (1980) Global atmospheric carbon dioxide measurements: a review of methodologies, existing programmes and available data. *Technical Report*, World Meteorological Organization, Geneva, Switzerland.

Revelle, R., and Munk, W. (1977) The carbon dioxide cycle and the biosphere. In: *Energy and Climate*, 140–158. National Academy of Sciences, Washington, DC.

Rodin, L. E., Bazilevich, N. I., and Rozov, N. N. (1975) Productivity of the world's main ecosystems. In: *Productivity of World Ecosystems*, 13–26. National Academy of Sciences, Washington, DC.

Rotty, R. (1982) Uncertainties associated with global effects of atmospheric carbon dioxide. ORAU/IEA-79-60(0), Oak Ridge Associated Univefsities, Institute for Energy Analysis, Oak Ridge, Tennessee.

SCEP (1970) Man's impact on the global environment: assessment and recommendations for action. In: *Study of Critical Environmental Problems*. M.I.T. Press, Cambridge.

Schlesinger, W. H. (1977) Carbon balance in terrestrial detritus. *Annual Review of Ecology and Systematics*, **8**, 51–81.

Whittaker, R. H., and Likens, G. E. (1973) Carbon in the biota. In: Woodwell, G. M. and Pecan, E. V. (eds), *Carbon and the Biosphere*. *AEC Symposium Series*, **30**, 281–302. Technical Information Center, Springfield, Virginia.

Wong, C. S. (1978) Atmospheric input of carbon dioxide from burning wood. *Science*, **200**, 197–200.

Woodwell, G. M. (1983) Biotic effects on the concentration of atmospheric carbon dioxide: A review and projection. National Academy of Sciences.

Woodwell, G. M., Hobbie, J. E., Houghton, R. A., Melillo, J. M., Moore, B., Park, A. B., Peterson, B. J., Shaver, G. R., and Stone, T. A. (1983a) Deforestation measured by LANDSAT. Report to the Department of Energy, Washington, DC. TR005.

Woodwell, G. M., Hobbie, J. E., Houghton, R. A., Melillo, J. M., Moore, B., Peterson, B. J., and Shaver, G. R. (1983b). The contribution of global deforestation to atmospheric carbon dioxide. In press, Science.

Woodwell, G. M., Houghton, R. A., and Tempel, N. R. (1973) Atmospheric CO_2 at Brookhaven, Long Island, New York: Patterns of variation up to 125 meters. *J. Geophys. Res.*, **78**, 932–940.

Woodwell, G. M., and Whittaker, R. H. (1968) Primary production in terrestrial ecosystems. *American Zoologist*, **8**, 19–30.

Woodwell, G. M., Whittaker, R. H., Reiners, W. A., Likens, G. E., Delwiche, C. C., and Botkin, D. B. (1978) The biota and the world carbon budget. *Science*, **199**, 141–146.

SECTION II

Vegetation

The Role of Terrestrial Vegetation in the Global Carbon Cycle:
Measurement by Remote Sensing
Edited by G. M. Woodwell
© 1984 SCOPE. Published by John Wiley & Sons Ltd

CHAPTER 2
Classification and Mapping of Plant Communities: a Review with Emphasis on Tropical Vegetation

D. MUELLER-DOMBOIS

Deparrment of Botany, University of Hawaii at Manoa, Honolulu, Hawaii

ABSTRACT

Methods of classifying vegetation are reviewed. Classification is aimed at portraying either potential or existing vegetation. Potential vegetation is commonly mapped using bioclimatic classification schemes following scholars such as Holdridge or Walter or landscape classification schemes such as those of UNESCO and Ellenberg. Maps of potential vegetation obtained from bioclimatic parameters can be used as predictors of primary production. Existing vegetation is mapped by either a classification based on physiognomy or on floristics. The physiognomic method provides a better estimate of phytomass because it takes into account variations due to succession and habitat. Bioclimatic and landscape classifications work equally well in temperate and tropical zones when mapped at a small scale. Problems in large-scale mapping of the tropics arise from the diversity of the species, different patterns of distribution, and lack of field studies. The author suggests mapping the existing vegetation of the tropics by combining satellite imagery with ancillary information from large-scale maps. Satellite imagery can be used to monitor the rate of loss of tropical ecosystems by use of successive images.

2.1 INTRODUCTION

The general objectives for classifying and mapping vegetation are three:

(1) Recognition and outlining of vegetation patterns for purposes of overview or inventory.
(2) Extrapolation of field observations and measurements to an appropriate level of geographic and ecological generalization.
(3) Explanation of vegetation patterns in terms of environment, past as well as present.

2.2 CLASSIFICATION SYSTEMS: POSSIBILITIES AND LIMITATIONS

The classification and the mapping of vegetation are fundamental tools for obtaining knowledge about the earth's vegetation cover and its relationship to the earth's environment. Classification has a twofold purpose: to sort the various patterns of plant communities that form a matrix of global, regional and local vegetation covers, and to investigate and explain their ecological relationships.

Two basic aspects of vegetation require explanation. They are its variation in space and its variation in time. One of the important tools for determining vegetation's variation in space is mapping. One of the important tools for determining variation in time is repeated mapping. This latter approach to vegetation dynamics has not been used as widely as it could have been, but the approach will gain importance in the future because of developments in aerial photography. This approach will also be promoted through the increased use of remote sensing techniques.

In addition to mapping, there are many other traditional and modern tools for documenting spatial and temporal variations in vegetation. These include profile diagrams (Davis and Richards, 1933, 1934; Beard, 1946, 1978), ordination diagrams (Whittaker, 1978a), dendrographs, two-way tabulations (Braun-Blanquet, 1928, 1965; Mueller-Dombois and Ellenberg, 1974), and many other multivariate analysis techniques (Orloci, 1978). These data-display tools differ in their degree of abstraction and are often complementary. Maps and profile diagrams are perhaps the most fundamental and easy to understand.

Classification requires two levels of abstraction, which are also transferred to most vegetation maps. The first level occurs from the process of recording a vegetation sample (or relevé)* in the field. In a vegetation sample, the researcher 'abstracts' a number of characteristics from a segment of the plant community of which the sample should be representative. This first-level abstraction applies also to vegetation descriptions, which are done without taking formal samples. The second level of abstraction relates to the grouping of individual samples into community types.

Both levels of abstraction have received a great deal of attention, and disagreement over the second level caused a major schism. Gleason (1926) believed that community samples differ so much from place to place that it was not appropriate to group them into classes or types. Three decades later a multidimensional ordination technique was developed by Bray and Curtis (1957) which was believed to do more justice to the multivariate nature of vegetation than classification does. More recently, the two viewpoints have been reconciled (Mueller-Dombois and Ellenberg, 1974; Orloci, 1978;

---

* The French word for 'abstract'.

Whittaker, 1978a), and the complementary aspects of the two techniques have been emphasized.

It is well to remember, however, that some detailed information is lost through the use of either process as certain generalizations are formulated. The value of a plant community classification lies in the formulation of generalizations which are the most appropriate ones for the purposes at hand. All classifications and all vegetation maps are purposive in nature, but they range from classifications and maps done for general ecological purposes to those done for very restricted purposes. In the latter group, for example, are many maps prepared for the exploitation of forests.

When existing maps and classifications are used for new purposes, such as answering questions about the global carbon dioxide cycle, they may not be adequate. New classifications or maps may have to be developed, or existing ones reinterpreted.

2.2.1 Criteria for Classifying Plant Communities

Vegetation is classified according to a number of criteria based on either the vegetation itself, the environment surrounding the vegetation or a combination of the vegetation and its environment.

Table 2.1 summarizes these criteria as physiognomic, floristic, environmental, geographic, successional and vegetation–environmental (two types).

The physiognomic criteria refer to the morphological patterns or form-variations of the vegetation cover. The basic plant community unit is therefore called a *formation*, and classification schemes using physiognomic criteria are often called formation systems. The subcriteria in Table 2.1 indicate that one can classify vegetation physiognomically by the dominant growth form or life form,* by a combination of life forms, by the characteristics of the vegetation layers (often easily distinguished by height stratification in forest communities) and by periodicity—for example, by the synchronized leaf fall and resprouting that occurs in certain forest and scrub communities. Analogous behaviour is the periodic dying of annuals in annual grasslands or the periodic drying and regreening of grass shoots in many perennial grasslands. The concept can be further extended to periodicity in flowering and fruiting and to the patterning of these phenomena within and between plant communities.

Floristic criteria are used for classifying and mapping variations in the distribution and composition of species in a vegetation cover. Depending on the vegetation and the purposes of the classification, one can use single

---

* Some authors (e.g. Beard, 1978) prefer to apply the term 'growth form' to trees, shrubs, grasses, etc. and the term 'life form' to the Raunkiaerian life forms (Raunkiaer, 1937), such as phanerophytes, chamaephytes, hemicryptophytes, etc. The latter term has a more functional connotation, but the distinction is not important here.

Table 2.1   Criteria for classifying vegetation

Properties of the vegetation itself	Properties outside the vegetation	Properties combining vegetation and environment

1 Physiognomic
  Dominant growth
  form or life form
  Combination of
  growth or life forms
  Layers
  Periodicity
 	
2 Floristic
  Dominant species
  Combination of
  dominants
  Certain groups of
  species:
    defined through
    tabulation
    techniques or
    numerical criteria
    (multi-variate
    analyses)
	3 Environment
	  (or habitat)
	  By individual
	    components
	    Climate
	    Topography
	    Land form
	    Soil
	    Human
	    influences
	  By combination of
	    site factors
	
	4 Geographical
	  location
	
	5 Successional stage
	  By presumed final
	    stage (climax):
	    defined by life-
	    form
	    combinations
	    or floristic
	    criteria or by
	    environmental
	    factors
		6 By independent analysis of
		  strictly vegetational criteria
		  and independent analysis of
		  environmental components
		  and subsequent correlation
		  (e.g. through map overlays)
		
		7 By combined analysis of
		  vegetation and environment
		  into integrated units:
		    Ecosystem classification
		    Emphasis is on functional
		    interrelationships

dominant species, combinations of dominants, or other groups of species. Community types classified through the use of a single dominant species or a combination of dominants have been distinguished by Whittaker (1962, 1978b) as *dominance types*. Communities classified through the use of species groups derived by means of the Braun-Blanquet tabulation technique, or species groups derived through mathematical multivariate techniques, have been distinguished as *association types* (Mueller-Dombois and Ellenberg, 1974; Whittaker, 1978b). Species or groups of species which are used to distinguish association types may also be dominant species, but the main criterion is that they must have restricted ranges or amplitudes within the vegetation cover to be analysed so that subdivisions or patterns can be distinguished. They can thus also be called differentiating (or differential) species.

Environmental criteria have been used for both formal and informal classification systems. Formal classification systems based on environmental criteria are, for example, the bioclimatic systems of Köppen (1936),

Thornthwaite (1948) and Holdridge (1967). Walter (1955, 1971, 1973a) has demonstrated a more flexible approach through the use of climate diagrams, which do not necessarily prescribe the boundaries of the areas to be classified. Various landscape classifications have been developed through the use of one or more of the other environmental criteria listed in Table 2.1 with the aim of predicting the vegetation potential of different segments of land.

Geographic criteria are often used in comprehensive classifications of vegetation on whole continents, such as those for the forests of South America by Hueck (1966) or the vegetation of Africa by Knapp (1973). These treatments are informal and do not follow a prescribed scheme of classification.

A highly formalized scheme for classifying successional stages, based primarily on environmental criteria and dominant species, was developed by Clements (1916, 1928).

Schemes based on independent analysis of strictly vegetational criteria are those of Dansereau (1957), Fosberg (1961, 1967), Küchler (1964) and Specht (1970). The ecological meaning of these classifications is determined by matching them against independently derived environmental classifications or maps.

Criteria that involve the combination of information about vegetation and environment integrated into units include the ecological series approach of Sukachev (1928), the biogeoclimatic zonation scheme of Krajina (1965, 1969) and the world ecosystem classification of Ellenberg (1973).

2.2.2 The Classic Systems and Their Limitations

The best known classic systems are:

(1) The physiognomic classifications of Grisebach (1872) and Drude (1902), which use only vegetation form without reference to species.
(2) The environmentally oriented classifications of Warming (1909), Graebner (1925) and Sukachev (1932), which use environmental factors as the primary criteria for plant community recognition.
(3) The physiognomic–environmental classifications of Schimper (1898), Diels and Mattick (1908), Brockmann-Jerosch and Rübel (1912), Du Rietz (1921) and Rübel (1933). (Du Rietz also used dominant species.) These systems use physiognomic criteria mixed with environmental descriptors.
(4) The areal–geographic–floristic classification of Schmid (1963), which uses the geographic distribution of species to define floristic provinces at various levels of detail.
(5) The dynamic–floristic classifications of Clements (1916, 1928), Tansley and Chipp (1926) and other American and British ecologists, which are based mainly on the final stages of succession (climax). The detailed classification system proposed by Clements was based on a number of successional

stages inferred from environmental and dominant species criteria. This system, however, is no longer used as a formalized approach.

(6) The floristic–structural classifications of Cajander (1909) and Braun-Blanquet (1928). Both use species groups with restricted amplitudes for defining vegetation units. Cajander's method is oriented toward the identification of forest site types, whereas Braun-Blanquet's approach is oriented toward analysing all kinds of mosaic patterns of species distribution or composition in a regional or local vegetation cover.

The physiognomic–environmental concept, for example, provided the framework for the International Biological Program (IBP) Biome studies in the US, in which the focus of research was on the functional interrelationships of the components of ecosystems (criterion 7 in Table 2.1). The physiognomic–environmental concept also influenced a classification system (Mueller-Dombois and Ellenberg, 1974) developed for UNESCO (1973) for mapping the world's vegetation at a scale of 1:1 million. A further step toward a scheme for classifying ecosystems according to criterion 7 was provided by Ellenberg (1973).

The successional viewpoint (criterion 5) has given rise to studies of the long-term dynamics of vegetation. For example, it has inspired ecological approaches to analyses of stand structure for predicting the future development of forest stands (Leak, 1965; Goff and West, 1975). The environmental viewpoint (criterion 3) has spawned modern approaches to environmental, or 'direct', gradient analysis (Whittaker, 1978a). The floristic approach has given rise to new methods of mosaic analysis called 'indirect' gradient analysis by Whittaker (1978a). The physiognomic approach has been successfully subjected to detailed numerical analyses (Knight and Loucks, 1969; Webb *et al.*, 1970, 1976) and used to develop new approaches for classifying vegetation biomass (Fosberg, 1967; Specht, 1970; Specht *et al.*, 1974).

The six classic approaches and their contemporary modification all have certain limitations. The original physiognomic classifications, for example, gave only broad global overviews of the world's vegetation cover. When refined strictly by physiognomic–structural criteria—for example, by considering finer categories in stature and plant spacing—they become artificial. When a forest is distinguished from a woodland simply by the stature and spacing of trees, such categorization may separate units of vegetation that are closely related in ecological terms. They may differ only in the woodland being a recently disturbed portion of the forest.

One way out of this problem is the formation series approach of Beard (1978), which employs an ecological series (in this case, a climatic gradient) to portray physiognomic responses in the stature of certain types of formations through generalized profile diagrams.

Such generalized profile diagrams, which have been used by many authors, are useful tools for illustrating regional relationships between physiognomy and environment. It would be questionable, however, to try to draw a single generalized profile for all tropical plant formations regardless of regional variations, just as it would be questionable to try to draw a single generalized profile for all temperate plant formations. For example, the redwood (*Sequoia*) forests of northern California and the Douglas-fir (*Pseudotsuga menziesii*) forests of the Pacific Northwest form ecosystems whose physiognomy is unique to North America. The major physiognomic characteristics—i.e., stature, spacing, crown shapes, leaf characteristics (including deciduousness versus evergreeness), layering and plant biomass relationships—could hardly be presented on a single global profile diagram. The main reason for this is that stature, crown shape and leaf characteristics are not merely a function of the environment but also of the floristic history and plant types in a particular region (Walter, 1971, 1976). Any attempt at a globally integrated physiognomic–environmental classification scheme (classic or contemporary) cannot really do justice to the variability of the world's vegetation cover.

The environmentally oriented classifications give even less information about the vegetation itself. They only define certain climatically, topographically or edaphically homogeneous segments of land. They can broadly predict, however, what sort of vegetation formation—and, sometimes, species composition—will occur in a given area, provided there are no interfering factors. Environmental gradient analysis is a powerful ecological tool that evolved from this approach.

The areal–geographic–floristic approach is concerned with the provinciality of plant species and communities. It also includes historical investigations of plant species and communities, which are important in explaining the variability of the vegetation cover.

The provinciality of plant species becomes a problem in the five other approaches to classifying vegetation. This was first realized during work on the floristic schemes. The Braun-Blanquet association scheme, which dealt with vegetation in central Europe, was originally thought to be transferable to other continents. It was found, however, that the key species useful for determining association types in one region could not always be used when an attempt was made to extrapolate the scheme over a broader geographic range. Certain key species disappeared, while others lost their indicator value over larger areas.

The latter phenomenon is related to the frequently documented observation that widely distributed species have geographic varieties (ecotypes) which have evolved different adaptations to their environments. Thus, the floristic association system is useful only for intensive studies of regional vegetation covers, and the community types defined in these regional studies cannot be expected to be the same in other regions with similar environmental relationships. This does not mean, however, that the floristic association system does not work in regions outside central Europe. It has been found to work in many other regions, and has even wider applicability as a research tool than classifications based on dominance types.

Floristic dominance-type classifications are useful only where plant communities can be identified by dominant species. This restricts their application in regions where floristic diversity is great. That includes most tropical forests and many subtropical areas.

Another difficulty with the dominance-type concept in central Europe led to the association-type approach. This was the small number of dominant tree species and their indifference to changes in habitat. The few dominant tree species, such as the European beech (*Fagus sylvatica*) and pine (*Pinus sylvestris*), range over broad areas with considerable environmental heterogeneity. In these areas undergrowth species often show a much closer relationship to variations in habitat.

It is also important to realize, however, that dominance types can be found in some parts of the tropics—for example, the *Mora excelsa* forest in Trinidad, the *Metrosideros polymorpha* rain forest in Hawaii, and the teakwood (*Tectonoa grandis*) and sal (*Shorea robusta*) forests in India. Wherever the existence of such dominance types conveys useful ecological information, a dominance-type scheme is superior because of its simplicity.

Because of the limitations inherent in all of the major classifications so far devised, it is probably not possible to devise a general purpose classification of the world's vegetation.

Once the limitations in each system are realized, however, it is possible to combine the aspects of each system which are best suited for such new purposes as mapping the plant biomass and primary production of the globe.

2.3　VEGETATION MAPS

Vegetation has been mapped at all geographic scales, ranging from global overviews at very small scales (such as 1:100 million) to individual research plots and one-metre square quadrats at very large scales (such as 1:10, where 1 cm on the map corresponds to 1 dm on the ground). It is helpful to distinguish among three categories of vegetation maps: small scale, intermediate scale and large scale.

2.3.1　Small-scale Maps

Such maps may be defined as ranging from a scale of 1:1 million (where 1 cm on the map represents 10 km in the field) to scales of much smaller size, such as 1:100 million. It is useful to distinguish within this range the very small-scale maps (from 1:10 million to 1:100 million) which are commonly used for global overviews.

The Brockmann-Jerosch map reproduced by Woodwell (1978: 38–39) is a very small-scale vegetation map on which 1 cm corresponds to approximately 1000 km on the earth's surface (i.e., 1:100 million). The map shows 10 world vegetation types, from tropical rain forest to tundra. These 10 were supplied by Woodwell (1978) with plant biomass data in kg C/m^2. The map uses the physiognomic–environmental scheme of vegetation classification.

Whittaker (1970, 1975) recognized 25 such types for the world. To accommodate Whittaker's 25 types on a global map the scale would have to be enlarged, perhaps to twice that of the Brockmann-Jerosch map, i.e., to a scale of 1:50 million. At this scale a world map is approximately the size of two large atlas pages. Thus, 25 physiognomic–environmental types can still be accommodated on a very small-scale map.

Perhaps the best contemporary vegetation map in the very small-scale range is the 1:25 million map of Schmithüsen (1968), which shows 144 physiognomic–environmental types and includes such important types as mangrove forests and tropical swamp forests, which could not be shown on the Brockmann-Jerosch map. It is of interest to note that Schmithüsen's map includes some dominance types. These are indicated by printed symbols and include such species as *Quercus ilex*, *Q. suber* and some regionally dominant tree genera, such as *Pinus*, *Sequoia* and *Podocarpus*. The 1:25 million scale is about the same scale as that used in a standard wall map of the world. Schmithüsen's map is reproduced as a small pamphlet with 11 double pages.

The differences among these three maps can only be discussed in relation to particular objectives. As an overview of plant biomass relationships, the Brockmann-Jerosch map of 10 vegetation types serves very well. Greater detail and perhaps accuracy would be obtained, however, by supplying plant biomass values for the 144 vegetation types mapped by Schmithüsen and then calculating the surface area of each type. Such an approach was used by Rodin *et al.* (1975), who supplied plant biomass and productivity data for 106 terrestrial soil-plant formations. Since these Russian authors used a larger map-scale with many more vegetation types, one may conclude that their global estimates are more accurate than the values shown by Whittaker and Likens (cited in Lieth and Whittaker, 1975: 13). The latter based their global estimate of primary production on 15 terrestrial vegetation types and give 164×10^9 tons (t) of dry matter per year, while Rodin *et al.* give 172×10^9 t. Lieth (1973) estimated 100.2×10^9 t, based on 20 terrestrial vegetation types. For standing plant biomass (in dry weight), Whittaker and Likens (1975) give 1837×10^9 t. Rodin *et al.* (1975) give 2402.5×10^9 t. These two, and different, summation values were calculated for the same continental land area of 149×10^6 km^2.

Of course, the accuracy of these estimates is dependent not only on the map scale and number of vegetation types used but also on an estimate of the variation within each vegetation type and on the number of actual measurements. Both variables have to be considered in arriving at a best estimate, or properly balanced mean value, for each vegetation type.

A better estimate of structural variation can be made through the use of existing vegetation maps at successively larger scales. The next group larger in scale than global vegetation maps are maps of individual continents. These range in scale from 1:1 million to 1:10 million and are thus still small-scale vegetation maps. Two examples are the vegetation map of the United States by Küchler (1965) and the vegetation map of South American by Hueck and Seibert (1972).

Küchler's pocket map is at a scale of 1:7.5 million, i.e., 1 cm represents 75 km. A comparison of the information content of Küchler's map with that given for the conterminous United States on the Brockmann-Jerosch map is shown in Table 2.2. The global map of Brockmann-Jerosch shows seven

Table 2.2   Comparison of vegetation units within the small map-scale range for the US: temperature and subtropical

Global vegetation map of Brockmann-Jerosch (1:100 million)	Continental vegetation map of Küchler (1:7.5 million). Units grouped hierarchically	
1 Boreal and montane forest (in the Pacific NW)	1 Western needleleaf forests (21 types)	8 Central and eastern grasslands and forest combinations (12 types)
2 Evergreen hardwood forest (in parts of California)	2 Western broadleaf forests (2 types)	9 Eastern needleleaf forests (5 types)
3 Steppe and prairie (in central US)	3 Western mixed broadleaf/needleleaf forests (5 types)	10 Eastern broadleaf forests (8 types, including a mangrove type)
4 An undesignated unit (probably also grassland in south-central US)	4 Western shrubland (11 types, including a desert type)	11 Eastern mixed broadleaf/needleleaf forests (10 types)
5 Dry desert and semi-desert (in south and southwestern US)	5 Western grassland (9 types)	
6 Summer-green deciduous forest (for most of eastern US)	6 Western shrub- and grassland combinations (7 types)	Total 106 types
7 Temperate rain forest = Laurel forest (for Florida)	7 Central and eastern grassland (16 types)	

general physiognomic vegetation types for the conterminous United States, while Küchler's map shows 11 geographic–physiognomic units which serve as a broad framework for grouping the actual map units, which are 106 major types of vegetation. In most cases, these are recognized by a combination of physiognomic criteria and dominant species. Thus, by enlarging the map scale by a factor of approximately 10, the number of mappable vegetation types can be increased by a factor of about 10 to 20.

This illustrates an important point which is often overlooked. The definition of plant communities or vegetation types is strongly dependent on map scale. Another point, and one that is relevant to the carbon dioxide question, is this: hidden in the global physiognomic vegetation types, which are commonly called biome types (Whittaker, 1970, 1975), is a great deal of structural 'noise'. This 'noise', however, is not random. It is the product of the mosaic of vegetation types.

Table 2.3 presents a comparison of vegetation types for tropical South America. The increase in scale from the Brockmann-Jerosch map to the map

Table 2.3   Comparison of vegetation units within the small map-scale range for tropical South America. (Excluding Central America and Temperate South America)

Global vegetation Map of Brockmann-Jerosch (1:100 million)	Continental vegetation map of Hueck and Seibert (1:8 million). Units grouped hierarchically where possible	
1 Equatorial and tropical rain forest (Amazon area)	1 Evergreen tropical rain forests (13 types in Amazon and Orinoco areas)	6 Tropical raingreen mesophytic forests (3 types)
2 Raingreen forest, woodland scrub and savannah (N and S of Amazon area)	2 Várzea swamp forest (Amazon River)	7 Savannahs and palm savannahs (5 types)
3 Temperate rain forest = Laurel forest (here seen as equivalent to tropical montane forest, Andes and parts of Brazil)	3 Campinas = grassland and open woodlands in rain forest area (Amazon)	8 Tropical dry forests (9 types)
4 Steppe and prairie grasslands (Venezuela)	4 Evergreen tropical montane forests (3 types in Andes, 2 types in Pacific and Caribbean region, 3 types in Atlantic region)	9 Tropical scrub and grassland (6 types)
5 Dry desert and semi-desert (Peru, Equador)		10 Desert and semi-deserts (4 types)
6 Highland areas with alpine tundra (tropical part of Andean Mountains)	5 Mixed evergreen/ raingreen tropical montane forests (3 types, Andes)	11 Tropical high-altitude vegetation (4 types) Total 57 types

of Hueck and Seibert (1972) is very similar to the previous comparison. The differentiation of vegetation types for tropical South America is increased by a factor of almost 10, i.e., from 6 to 57 types. If Central America had been included, the increased differentiation of vegetation types associated with the increase in map scale would have been even more comparable. The Hueck–Seibert map shows that tropical vegetation is just as varied as temperate vegetation, a point emphasized by Whitmore (1975) and Poore (1978). Many of the vegetation types on Hueck and Seibert's map are identified by colloquial names, such as Várzea forest, Alisio forest, Caatinga beixa, Campos Cerrados and Campos Limpos. These terms convey a particular physiognomy only to those familiar with South American tropical vegetation. It would not be difficult to translate them into internationally understandable terminology of vegetation structure for use in geographic calculations of plant biomass and primary production.

However, all small-scale vegetation maps have a major defect with respect to calculating plant biomass. They do not portray the outlines of existing vegetation. Küchler (1965) called his map a map of 'potential natural vegetation'. In other words, his map outlines certain areas where such vegetation could grow. Small-scale vegetation maps are not, strictly speaking, vegetation maps. They are site maps. They may be very useful for calculating primary production, which is more a function of site factors than plant biomass, but they may give erroneous values for biomass if an attempt is made to extrapolate real vegetation from potential vegetation.

2.3.2 Intermediate-scale Maps

These may be defined as ranging from 1:1 million to 1:100 000 in scale. The upper range of this scale accommodates 10 km, and the lower 1 km, on 1 cm of map. An example of an intermediate-scale map is Küchler's vegetation map of Kansas (1974). This map was published at a scale of 1:800 000 (1 cm = 8 km). It illustrates how much more information on vegetation can be displayed when a section of a continental map is enlarged by a factor of approximately 10. There is good similarity between the vegetation in the 4.5 × 9 cm rectangle that is Kansas on Küchler's map of the United States and the 40 × 80 cm vegetation map of Kansas alone. However, the number of vegetation types on the latter is 16, compared to 7 on the former.*

Another map in the same scale range is the vegetation map of the Hawaiian Islands prepared by Ripperton and Hosaka (1942). The island of Hawaii is portrayed in Figure 2.1 at a scale of 1:1.8 million (1 cm = 18 km). This map

---

* Küchler's map of Kansas is called a map of potential natural vegetation. This map should not be confused with a map showing the outlines of existing vegetation types. One cannot deduce existing plant biomass easily from this map, but it may serve very well for providing indices of primary production.

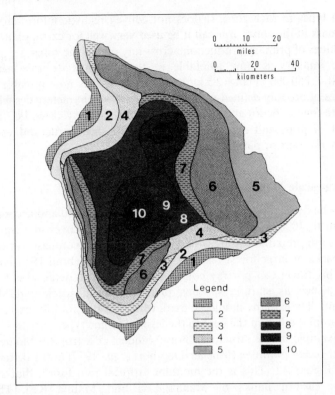

Figure 2.1 Vegetation zone map of the Island of Hawaii originally at 1:1 million (from Ripperton and Hosaka, 1942). Zone names modified: (1) Savannah and dry grassland, (2) Dry sclerophyll forest or scrub, (3) Mixed mesophytic woodland or scrub, (4) Mixed mesophytic forest and savannah, (5) Lowland rain forest, (6) Montane rain forest, (7) Upper montane rain forest, (8) Mountain parkland and savannah, (9) Subalpine forest and scrub, (10) Sparse alpine scrub and stone desert

shows 10 vegetation zones, rather than vegetation types, and that makes interpretation somewhat easier. The boundaries of the vegetation zones are the result of several factors, of which temperature and rainfall are the most important. Vegetation is also included as an indicator, but soil and substrate are largely ignored. This is apparent from the fact that there are a number of recent lava rock substrates that extend south and southeast from the top of Mauna Loa to the coast and northeast through the saddle between Mauna Loa and Mauna Kea toward Hilo. Many of the lava deposits to the south are practically devoid of vegetation, and it will take more than 2000 years for the deposits in the coastal lowland to become vegetated with anything resembling the plant biomass on adajacent soils. Thus, the map is not a map of potential vegetation. It is a bioclimatic map that includes a mosaic of different structural vegetation types in each zone. It does not convey realistic information about plant biomass in the zones, nor can it be used very well for extrapolating single average values of primary production across any one of the zones.

Satellite imagery is now available in the intermediate-scale range, for example at 1:250 000 (1 cm = 2.5 km). At this scale it is now possible to map the outlines of broadly defined vegetation types. Two characteristics which can be used are tone (albedo) and texture. These can, in many cases, be translated into forest or grassland or other very broadly defined structural vegetation units, such as desert or parkland (Reeves *et al.*, 1975).

2.3.3 Large-scale Maps

These may be defined as ranging from 1:100 000 to 1:10 000 and accommodate from 100 m to 1000 m of ground surface or vegetation cover in 1 cm of map. Such maps are prepared for areas of county size that encompass, for example, national parks or large forest reserves, that is, areas of about 10^2 to 10^4 km^2. Maps in this range can portray considerable structural detail about existing vegetation, such as stature, spacing, plant life form, species and biomass relationships. These maps should be used, whenever possible, for extrapolating biomass samples taken in the field to the level of biome type.

As an example of the structural 'noise' hidden in a tropical biome type, I refer to our recent mapping (Mueller-Dombois *et al.*, 1977) of the better part of zone 6 in Figure 2.1. This is the montane tropical rain forest, that stretches across the eastern flanks of Mauna Loa and Mauna Kea. The area encompassed by our large-scale map is approximately 80 000 ha (16 × 50 km). The map was prepared from colour aerial photographs at a scale of 1:12 000 and reduced to a scale of 1:48 000. On this map are shown five major structural types:

(1) Dense 'ōhi'a (*Metrosideros polymorpha*) forests with crown cover of more than 85 per cent	(2 subtypes)
(2) Closed 'ōhi'a forests with crown cover from 60 to 85 per cent	(5 subtypes)
(3) Open 'ōhi'a forests with crown cover from 15 to 60 per cent	(7 subtypes)
(4) Low-growing vegetation with scattered trees, less than 15 per cent cover	(5 subtypes)
(5) Low-growing vegetation with shrub-like trees under 5 m tall	(3 subtypes)
Total	22 subtypes

The subtypes here are the actual units on the map, and they are further distinguished by variations in stand stature, associated dominant species, and mappable characteristics of the undergrowth.

Large-scale vegetation maps have also been used as site indicator maps following research to establish correlations between certain vegetation patterns and the physical characteristics of the sites. Several examples are given in Mueller-Dombois and Ellenberg (1974: 411 ff.). Such maps can be used to project primary production values for local areas. Webb (1968) has worked out such site indicator properties for tropical rain forests in Queensland.

Unfortunately, published large-scale maps are few (Küchler, 1966, 1968, 1970; Küchler and McCormick, 1965). Some temperate areas, particularly western Europe, are well covered by published vegetation maps, but few maps of tropical vegetation exist. A number of such maps are in preparation for southeast Asia, according to de Rosayro (1974).

2.4 USE OF CONTEMPORARY CLASSIFICATION SYSTEMS IN VEGETATION MAPPING

Any ecologist who has done a certain amount of mapping of the vegetation will have made at least three observations relevant to mapping and classification of plant communities. The first is that certain vegetation boundaries are easy to draw because they are virtually self-evident on aerial photos or in the field, but that other boundaries must be drawn rather arbitrarily between two centres of structural or species compositional changes. The difficulty relates to the much-discussed continuum problem in vegetation (see Dansereau *et al.*, 1968). This problem is not unique to vegetation, however. It applies to all landscape components, such as topography, soils and geological formations.

A second observation is that difficulty is often encountered in integrating vegetation map units across a larger terrain. A new map unit may fall between two recognized unit categories. This is the problem of the classificatory continuum (McIntosh, 1967; Mueller-Dombois and Ellenberg, 1974), which usually requires repeated adjustment during the mapping process.

The third observation is that plant community definition, and therefore classification, is really a matter of geographic scale. One can argue a great deal about the merits of a particular classification scheme, but once the geographic scale is given the argument is narrowed substantially.

Related to the problem of scale is the question of whether one can really map the outlines of existing plant communities. This depends on the type of ground-surface imagery available from remote sensing.

It is useful to separate contemporary classification systems into two general categories: those capable of mapping the potential vegetation of an area, and those capable of mapping its existing vegetation.

2.4.1　Schemes for Mapping Potential Vegetation

Potential vegetation refers to the inherent capacity of a land area to support a particular kind of vegetation. In the small map-scale range, potential vegetation can usually be mapped only in physiognomic categories, but at the larger map-scales one may be able to predict general species and tree growth capacities through site or habitat mapping (Daubenmire, 1968; Mueller-Dombois and Ellenberg, 1974). The usefulness of these methods, however, also depends on the floristic complexity of a particular region. The importance of this can be seen by comparing Küchler's (1965) map of the potential vegetation of the conterminous United States with Hueck and Seibert's map of the potential vegetation of the South American tropics (Tables 2.2 and 2.3). Hueck and Seibert use mostly colloquial physiognomic terms, while Küchler uses dominant species combinations for designation. The difference is due to the greater diversity of species in tropical than in temperate America.

Two contemporary schemes are used for mapping potential vegetation: bioclimatic and landscape. *Bioclimatic mapping* is used in the small- and intermediate-scale map range. Analogous approaches useful for large-scale mapping are *environmental gradient analysis* in the sense of Whittaker (1967) and the *scalar approach* of Loucks (1962).

The large-scale version of the landscape approach is also known as *habitat-type mapping* (Daubenmire, 1968; Mueller-Dombois, 1964), or as the *ecological series approach* (Sukachev, 1928; Mueller-Dombois, 1965), or as the *biogeocoenotic approach* (Sukachev, 1945; Krajina, 1960; Sukachev and Dylis, 1974) or as *ecological land classification* (Mueller-Dombois and Ellenberg, 1974: 319).

2.4.1.1 Bioclimatic Methods

The better known bioclimatic methods for mapping zones of vegetation at global or regional levels are those of Köppen (1931, 1936), Thornthwaite (1948), Thornthwaite and Mather (1957), Holdridge (1947, 1967) and Walter (1955, 1957, 1971, 1973a, 1973b, 1976). All use selected climatic characteristics that outline zones within which a certain general level of vegetation homogeneity should be found. They also suggest a strong similarity of vegetation in equivalent bioclimatic zones in different parts of the globe.

Köppen's approach utilizes five broad zones identified by capital letters: A, humid tropical climates; B, arid and semi-arid climates; C, temperate climates; D, cold continental climates; and E, periglacial climates. These broad zones are further subdivided by rainfall and temperature threshold values. For example, the humid tropical A climates are all defined as having monthly mean temperatures over 18 °C. They are subdivided into four:

> Af— tropical continuously wet (mean rainfall each month > 60 mm)
> Aw— tropical winter-dry (at least one month during the cooler season with rainfall under 60 mm)

As— tropical summer-dry (at least one month during the hotter season with rainfall under 60 mm)

Am—tropical monsoon (high annual rainfall with short dry season)

The method works very well as a first overview, but it includes such oddities as temperate-zone climates in tropical mountains. Apart from its rigidity, the system also suffers from a terminological problem.

Thornthwaite's system uses mean monthly rainfall and potential evapotranspiration. The latter value is derived through mean annual and monthly temperatures and a latitudinal correction factor, which takes seasonal changes in day length into consideration. The mean monthly potential evapotranspiration values, which are calculated from a formula, are plotted on a climate diagram together with the mean monthly rainfall values of a climatic station. The drying power of an environment is indicated on Thornthwaite's diagrams wherever the mean monthly rainfall curve falls below the potential evapotranspiration curve. The area on the diagram subtended by the two curves for the time period of such undercutting is subdivided into a period of soil water withdrawal, and the amount of this water is arbitrarily taken as 100 mm. When this storage water is used up, the remaining area between the curves is considered to reflect soil water deficiency or soil drought. The method makes it possible to estimate irrigation needs on arable cropland. It also works well to indicate the vegetation potential of an area through mapping of the diagrams and the use of certain threshold values for outlining bioclimatic zones. The method has not been tested widely in the tropics, where the calculation of potential evapotranspiration may require the incorporation of humidity in addition to temperature.

Holdridge's (1947, 1967) 'life zone' mapping method has become very popular in tropical America (Ewel and Whitmore, 1973; Barbour *et al.*, 1980). It employs two climatic characteristics, mean annual rainfall (or precipitation) and mean annual biotemperature. The latter is defined as the sum of mean monthly air temperatures (°C) divided by 12, with the provision that monthly temperatures of 0 °C and colder and those 30 °C or warmer are disregarded. For the tropics, mean annual biotemperatures are generally identical to mean annual air temperatures. A third characteristic, namely, a potential evapotranspiration ratio, is computed from the first two. Potential evapotranspiration is calculated from temperature by multiplying the annual biotemperature of a station (°C) by a factor of 58.93. A potential evapotranspiration ratio of one implies that water loss from evapotranspiration is equal to water gain from precipitation. These three climatic characteristics are worked into a triangular nomogram. Thirty world life zones are indicated in this triangular nomogram, each in form of a small hexagon. Life zones are outlined on maps simply by obtaining the mean annual precipitation and mean annual biotemperature from a number of climatic stations and by applying these data to the nomogram. One can also obtain the two characteristics from mean annual rainfall isohyets and calculation of the temperature lapse rate from stations at sea level. The nomogram informs the mapper of the threshold values for boundary interpolation and the life zone name.

Holdridge defines the tropical region as the global terrain with mean annual air temperatures of 24°C or greater. This is the narrowest proposed definition. Köppen's 18 °C is a more adequate index. Most authors, however, define the tropical belt as the geographical area between the Tropics of Cancer and Capricorn. Thus Holdridge's system, though providing an ingeniously derived climatic index, introduces a confusion in terminology. For example, according to Holdridge's system, Hawaii would be part of the subtropics because its mean annual temperature at sea level is near 23.5 °C. This delimitation is opposed to all conventional views. The natural lowland and lower montane vegetation in Hawaii is typically composed of tropical taxa and plant life forms. A second terminological problem is the naming of the life zones themselves. In his tropical belt (> 24 °C), for example, the sequence from wet to dry is rain forest, wet forest, moist forest, dry forest, very dry forest, thorn woodland, desert scrub and desert. These terms convey the impression of physiognomic vegetation units and thus prolong confusion between real vegetation and climate. It would be difficult, for example, to distinguish the rain forest category from the wet forest category physiognomically. None of the zones are physiognomic vegetation types. They are instead rigidly constructed bioclimatic zones. Beard (1978) has also commented on the rigidity and unfortunate terminology in the Holdridge system.

As a bioclimatic zonation scheme, however, Holdridge's system has an advantage in its simplicity of application. It may also be useful as a predictor of primary production. Since it is based on mean annual climatic characteristics, however, it lacks a clear definition of seasonality.

Walter's climate-diagram method has certain advantages over the preceding schemes even though, like them, it also uses only rainfall and temperature data. Like Thornthwaite's method, Walter's employs the annual sequence of mean monthly precipitation values of any given climatic station. These are plotted into the climate diagram.

In addition, Walter uses mean monthly temperature data without converting them into another form. The precipitation and temperature curves are plotted so that 10°C on the left-hand ordinate of each diagram corresponds to 20 mm of precipitation on the right-hand ordinate. The diagrams indicate a significant dry season (called drought season by Walter) wherever the precipitation curve falls below the temperature curve. The 10 °C/20 mm relationship is based on a suggestion by Gaussen and supported by empirical observations. It gives an index of environmental dryness, which is the third important characteristic in Walter's diagrams.

The method has been criticized because the mean monthly temperature curve does not take into account the varying relationship of potential evaporation to the temperature and humidity regime of a place. Of course, radiation and wind would be other important influences. A dry season in a tropical lowland environment as indicated on a climate diagram is, of course, drier in absolute terms than a similarly indicated dry season in a temperate climate. Moreover, a dry season in an arid tropical environment produces a proportionately greater evaporative power than a similar dry season in a humid tropical climate.

Walter answers this criticism with the equally logical observation that natural vegetation is acclimatized to the place in which it occurs. This implies that a short dry season in a humid tropical climate produces greater stress on indigenous vegetation than a short dry season on vegetation indigenious to a drier climate. Or, stated in the opposite way, a wet month (rainfall in excess of 100 mm) has a greater impact on dry-zone vegetation than it does on humid-zone vegetation. The 1/2 ratio of temperature to precipitation seems to have given a realistic drought index wherever the climate-diagram method has been tested, except in temperate climates with low winter precipitation. In these climates Walter has used as a second index a 1/3 relationship, which is indicated by a dashed line in the diagrams. The 1/3 ratio lowers the mean monthly rainfall curve relative to the temperature curve.

The climate-diagram method emphasizes many more climatic characteristics, where these are ecologically useful. In certain temperate environments, for example, the annual growing season is indicated by a line drawn at 10 °C, which makes it possible to estimate the average number of days above 10 °C. In some of the tropical climate diagrams the average day and night variations in temperature are shown. Each climate diagram also gives the annual mean precipitation and temperature at the station, which are the two basic kinds of data used in Holdridge's system. Thus, Walter's diagrams, though easy to understand, provide a great deal more climatic information than do other methods.

Walter's diagrams have been mapped by Walter and Lieth (1960) in a world atlas that shows climate diagrams for numerous places all over the globe. A more recent version in normal-sized book format (Walter *et al.*, 1975) contains eight maps, one for each continent or major world region. Each map is at the scale of 1:8 million and is filled with climate diagrams. These allow the user to appraise climatic variations over each continent and between continents with little effort. (The use of this method for identification of forest types on satellite imagery is discussed below.)

Unlike the other bioclimatic schemes discussed, Walter *et al.* (1975) do not prescribe boundaries for bioclimatic zones on the continental maps. Boundary designation is left to the user, who can quickly determine the climatic variation related to a distribution problem, for example, forest distribution on satellite imagery.

On a global map which accompanies the eight continental maps in the 1975 atlas, Walter outlines 10 world climatic zones. These correspond to traditional biome concepts, such as equatorial rain forest, tropical deciduous forest, boreal forest, etc., but it should be kept in mind that these are bioclimatic zones. They do not identify vegetation types. They are simply zones of a certain climatic character, and they always contain a variety of real vegetation types.

2.4.1.2 Landscape Classifications

In contrast to the bioclimatic methods, the landscape classification methods usually give greater emphasis to actual vegetation characteristics. They do this by integrating vegetation and certain aspects of the environment into ecosystem or landscape types. The 10 physiognomic–environmental world vegetation types mapped on the Brockmann-Jerosch map (in Woodwell, 1978) or the 104 soil-vegetation formations used by Rodin *et al.* (1975), which were used for presenting world primary production and phytomass values, are such landscape types.

Five landscape classification methods have proven their usefulness in different situations. Examples of these are the schemes of Krajina (1965, 1969), Hueck (1966), Hueck and Seibert (1972), UNESCO (1973), Ellenberg and Mueller-Dombois (1967a), Ellenberg (1973), Gaussen (1954, 1957, 1964) and Gaussen *et al.* (1964, 1965, 1967). These methods are particularly suitable for small and intermediate map-scales. I exclude discussion of landscape classifications particularly adapted to large-scale mapping.

Krajina's biogeoclimatic zonation scheme has been applied to a 1:1.9 million map of British Columbia (Krajina, 1969, 1974). In this classification system, plant community and soil types are integrated through detailed regional studies at the large map-scale range. Krajina calls these integrated units 'biogeocoenoses' in the tradition of Russian authors (Sukachev and Dylis, 1964). For outlining biogeoclimatic zones in the small map-scale range, he uses Köppen's bioclimatic scheme as a guide. However, the zonal boundaries are adjusted with the aid of information from soil and topographic maps. Each of Krajina's zonal units is characterized by a zonal soil type, zonal climate type and zonal vegetation type. The latter he defines as climatic climax vegetation, which is restricted to normally well drained soils on level to moderately varied (but not steep) topography. Mature vegetation on excessively or poorly drained soils is called an edaphic climax. Other variations include mature vegetation on alluvial flats or bogs, or vegetation on steep slopes. The latter are called topographic climax types. Krajina considers edaphic and topographic climax types to be significant parts of his zonal units. He therefore gives equal weight to these and to climatic climax vegetation in describing biogeoclimatic zones. Krajina's scheme therefore clearly recognizes the mosaic nature of vegetation types in the zonal map units.

Hueck's scheme for mapping the vegetation of South America at a scale of 1:8 million (Hueck and Seibert, 1972) is rather informal. Like Krajina's, it has been worked out by classifying vegetation from detailed observations in the field. Hueck's method is based on extrapolating the prevailing, existing natural and semi-natural vegetation across segments of the landscape which may now be under agriculture, urbanization or may be variously disturbed through fire and grazing. The extrapolation is based on an assessment of the total environment—i.e., climate, geology, soil, topography and history. The major tools used were existing maps.

It is interesting to compare, for example, Hueck and Seibert's map of Venezuela with the Holdridge map of Venezuela published at a scale of 1:2 million (Ewel *et al.*, 1968). The Hueck and Seibert map, although four times smaller in scale, contains more information about the vegetation. For example, it shows the ecologically important gallery forests along the Orinoco River; these are missing from the Holdridge map because it uses only mean annual rainfall and temperature relationships as a predictor. In short, important edaphic plant formations will not appear on bioclimatic maps but do appear on carefully done landscape maps.

The UNESCO system (1973) was developed for mapping the world's vegetation at a scale of 1:1 million (Ellenberg and Mueller-Dombois, 1967a). This project (Fontaine, 1978) involves mapping the world on map sheets, each of approximately state or province size. UNESCO set up an International Committee for Vegetation Mapping to develop an appropriate classification system, since none of the existing schemes was found to be suitable. A scheme based largely on work by Schmithüsen and Ellenberg was then developed. This scheme makes use of physiognomic units, which are further defined by environmental criteria. In principle, these units are similar to the Schimper (1898) units or the Brockmann-Jerosch and Rübel (1912) units shown by Woodwell (1978).

The system is hierarchical, with a top level consisting of five formation classes, as follows:

(1) Closed forests.
(2) Woodlands or open forests.
(3) Scrub or shrubland.
(4) Dwarf-scrub and related communities.
(5) Herbaceous vegetation.

Satellite imagery at the intermediate range may allow a mapper to outline such broad classes of existing vegetation.

At the second level, called the formation subclass, closed forests are subdivided into mainly evergreen, mainly deciduous and extremely xeromorphic forests.

The third level is called the formation group, which under 'closed forests mainly evergreen', for example, recognizes the tropical rain (or ombrophilous) forest. The tropical rain forest is then subdivided into eight actual formations:

(1) Tropical lowland rain forests.
(2) Tropical submontane rain forests.
(3) Tropical montane rain forests.
(4) Tropical subalpine rain forests.
(5) Tropical ombrophilous cloud forests.
(6) Tropical ombrophilous alluvial forests.
(7) Tropical ombrophilous swamp forests.
(8) Tropical evergreen bog forests.

These formations are further subdivided into subformations where useful. For example, the tropical montane rain forest is subdivided into a broad-leaved subformation (the most common), a needle leaved subformation, a microphyllous subformation, and a subformation rich in bamboo.

The UNESCO system is primarily a physiognomic system, but it also uses environmental terms for recognition. For example, lowland and submontane rain forests are separated by a topographic (climatic) boundary which has to be determined in each particular situation. In some tropical forest areas this boundary may correspond to a physiognomic or floristic segregation, but in others it may not. One can therefore also use this system in a landscape sense. For example, one may extrapolate a lowland rain forest as the typical physiognomic formation across a totally disturbed rain forest terrain.

Ellenberg's (1973) classification of world ecosystems brings a functional viewpoint to the UNESCO scheme. It begins with the biosphere as the largest ecosystem. This is subdivided into five mega-ecosystems:

M—Marine ecosystems (saline water as life medium).

L—Limnic ecosystems (fresh water as life medium).

S—Semi-terrestrial ecosystems (wet soil and air as life medium).

T—Terrestrial ecosystems (aerated soil and air as life medium).

U—Urban-industrial ecosystems (structures and creations of man as primary life medium).

The first four are considered natural or predominantly natural mega-ecosystems for which the sun provides the main energy source; the fifth is considered artificial and its energy sources are primarily fossil fuel or atomic power. In addition to life medium and energy source at the first level, Ellenberg incorporates five other functional criteria at the lower levels. These are:

(1) Biomass and productivity.
(2) Factors limiting the activity of primary producers, consumers and decomposers (e.g., seasonality).
(3) Regulating mechanisms of nutrient loss or gain (e.g., fire).
(4) Relative role of secondary producers (i.e., herbivores, carnivores, parasites and other mineralizers).
(5) The role of man (i.e., his role in the origin, development, energy flow and mineral cycling of the ecosystem, particularly in supplementing energy sources).

Four human roles are recognized in Ellenberg's scheme. These are:

(1) The harvesting of organic materials and minerals, which are significant for the metabolism of an ecosystem.

(2) The adding of minerals, organic materials or organisms.
(3) The adding of abnormal substances which are detrimental to important organisms or organism groups (toxification).
(4) The changing of species composition by suppressing existing species or by the introduction of alien species.

A rating scale of from one to nine is suggested for expressing the degree of each kind of human interferences, i.e., one (no toxification) to nine (excessive toxification).

The units below the mega-ecosystems are called macro-ecosystems. These are subdivided into meso-ecosystems, which in turn are divided into micro-ecosystems.

The macro-ecosystems are distinguished on the basis of biomass and productivity at a very general level: forest, grassland, desert, etc. Meso-ecosystems form the basic unit in Ellenberg's scheme and refer to such types as tropical evergeen rain forests with their animal life and cold-deciduous broad-leaf forests with their animal communities. Micro-ecosystems are such divisions as lowland, submontane, montane, etc.

The Ellenberg scheme also includes a biogeographic separation into nine regions, such as tropo-American, tropo-African, tropo-Asian, Australian, etc. Each of these biogeographic regions can be subdivided into biogeographic subregions or provinces. The system of biotic regions devised by Dasmann (1973) may be usefully applied here. Ellenberg's classification system has not been tested by actual mapping, but it should be similar to the UNESCO system, the only difference being that it adds significant functional information to a scheme which is primarily structural.

Gaussen's regional landscape system is perhaps the most complete of those discussed thus far. It has been applied successfully to south India (Gaussen *et al.*, 1961; Blasco, 1971) and Sri Lanka (Gaussen *et al.*, 1964, 1965). Gaussen's system involves the mapping of vegetation–landscape units at a scale of 1:1 million. Several of these maps have been produced at the French Institute in Pondichery, India.

These 1:1 million international vegetation maps cover a region the size of an individual state, province or large island (such as Sri Lanka) on a single fold-out sheet. The maps show the major vegetation zones, which correspond in level of detail to the biogeoclimatic zones of Krajina in British Colombia. For Sri Lanka, which is approximately 420 km long and 220 km wide in its widest part, Gaussen *et al.* (1964) recognize six vegetation zones plus a seventh narrow coastal zone of mangroves. The vegetation zones are named after two to four characteristic tree genera and are called vegetation series—for example, the *Manilkara–Chloroxylon* series, which is a xerophytic vegetation zone found in two places in Sri Lanka.

The zones are outlined on the basis of a bioclimatic analysis using the annual sequence of temperature and precipitation plotted at a ratio of 1/2. This bioclimatic analysis is combined with an analysis of prevailing physiognomic vegetation types and characteristic key species or genera. Major soil types at the soil order level are also included, but mapped independently on a smaller scale (1:5 million) inset map. Thus, the main criteria are synthesized from the vegetation and the climate (Mueller-Dombois, 1968). Gaussen's zones are therefore landscape units rather than physiognomic vegetation types.

Within each zone are shown a few generalized vegetation types which give some idea of the structural 'noise'. For the xerophytic *Manilkara–Chloroxylon* zones in Sri Lanka, for example, Gaussen *et al.* recognize two generalized or prevailing vegetation types, namely, semi-deciduous forest and scrub-woodland. These generalized physiognomic units can be outlined by satellite imagery. In this case they were outlined from standard aerial photographs or mosaics at a scale of 1:32000 and from already existing forest type maps (Gaussen *et al.*, 1965).

Gaussen's maps also show where natural vegetation has been removed and the prevailing use on converted lands. The land use is indicated by overprint symbols on a white background. White indicates total conversion.

In brief, Gaussen's maps provide a storehouse of ecological information on the natural resources of any region mapped by this landscape system. The maps show existing vegetation, vegetation potential, current uses of the land, climate, soils, topography, geology and administrative divisions. The only problem is that they are rather hard to read because they contain so much information. Map overlays would reduce the problem. Nonetheless, a user willing to spend some time looking over a Gaussen map can find as much information on it as in a thoroughly descriptive text. Indeed, so much integrated information is usually not found in a single book.

2.4.2 Methods for Mapping Existing Plant Communities

Methods concerned with characterizing and outlining existing plant communities differ from those concerned with potential vegetation in that the former extract mapping criteria solely from the structure of the vegetation. Environmental, historical or biotic factors, which may strongly determine the structure of a particular vegetation, are analysed independently. Thus, the focus is strictly on the pattern of the vegetation. Community patterns are established first, and the search for the causes of the patterns comes second. This approach (criterion 6 in Table 2.1) has certain advantages. For example, the procedure for mapping vegetation is relatively straightforward and thus much less time-consuming than the mapping of vegetation potential by landscape procedures. Direct vegetation mapping also avoids the pitfalls of circular reasoning.

For classification and mapping purposes it is useful to recognize both vegetation architecture and vegetation floristics. Vegetation architecture includes all recognizable forms of vegetation structure but usually excludes the floristic aspects. (The term 'vegetation structure' refers to all of the morphological characteristics of a vegetation, i.e., to its physiognomy and its species composition.) Vegetation floristics is primarily concerned with the distribution of plant species and species-population structure. A number of methods have been developed for classifying and mapping existing vegetation by using either architecture or floristics, or both.

2.4.2.1 Architecture of Vegetation

Classification schemes based on vegetation architecture use four structural criteria. These are (1) plant cover or spacing, (2) plant height or stature, (3) canopy and foliage characteristics and (4) plant growth form or life form. The difference between certain formalized schemes is simply that they use these four basic structural attributes in different sequences or combinations and with different subdivisions.

Four well-known examples of formalized architectural vegetation schemes are briefly characterized in the following discussion. Some related informal approaches are then mentioned. The four examples relate to the different schemes of Dansereau, Küchler, Fosberg and Specht. (The rather new and original approach to classifying tropical rain forest architecture pioneered by Webb (1968) and Webb *et al.* (1970) is discussed later under Problems and Measures in Mapping Tropical Vegetation.)

Dansereau's (1951, 1957) profile-diagram method offers a highly schematic and detailed approach to the mapping of vegetation architecture. Dansereau uses six categories for abstracting architectural characteristics from a given segment of vegetation in the field, namely, (1) plant life form, (2) plant size, (3) coverage, (4) foliage seasonality, (5) leaf shape and size and (6) leaf texture. Each of these six categories is further subdivided. For example, he recognizes six life-form types: trees, shrubs, herbs, bryoids, epiphytes and lianas. Each life-form type is represented by a symbol to show the plant type on a profile diagram. For example, a tree is symbolized by a circle above a vertical line (⊙), a shrub by a circle above an inverted triangle (⊽), a herbaceous life form by an inverted triangle alone (∇). The profile diagram is prepared in the field on graph or cross-section paper using eight spaces on the *y*-axis or height scale and 25 spaces on the *x*-axis or length scale for any vegetation segment. The *y*-axis shows the height of the life-form types in metres, while the *x*-axis shows the life forms in estimated quantities with appropriate spacing so as to convey the vertical crown or shoot projection as accurately as possible. Height and coverage are usually estimated, but they can be measured if necessary. Other structural characteristics are superimposed on the stylized symbols. For example, a deciduous tree-life form is represented by an open circle which symbolizes the crown. An evergreen tree-life form is indicated by cross-hatching of the circular crown symbol. A semi-deciduous tree-life form is indicated by vertical hatching. The same is done for shrub and herbaceous life forms, where applicable.

Leaf shapes are indicated similarly. For example, an inverted heart inside a circular crown symbol indicates a broad-leaved tree-life form. The use of such symbols makes Dansereau's profile diagrams somewhat difficult to understand. The method is straightforward, however, and when applied properly can convey considerable detail about the particular architecture of any given plant community sample.

Classification is done by sorting the profile diagram samples according to their similarity. Dansereau classifies vegetation from below, but he does not prescribe vegetation units or classes. Mapping is done by spatial extrapolation from a number of samples.

Küchler's (1947, 1949, 1967) formula method classifies vegetation architecture by using a hierarchical approach. It starts by dividing vegetation at the first level into two broad categories: 'basically woody' and 'basically herbaceous'. Küchler recognizes seven basically woody vegetation types, each with a letter symbol:

B = Broadleaved evergreen.
D = Broadleaved deciduous.
E = Needleleaved evergreen.
N = Needleleaved deciduous.
A = Aphyllous vegetation.
S = Semi-deciduous (B + D).
M = Mixed (D + E).

Three dominantly herbaceous vegetation types are also recognized:

G = Graminoid vegetation (all grassland and sedge communities).
F = Forb vegetation.
L = Lichen and moss cover.

These broad categories are then further distinguished by the dominance of specialized life forms within them, such as C = Climbers, K = Stem-succulents, T = Tuft plants (such as palms, cycads or tree ferns), V = Bamboos and X = Epiphytes.

A further subdivision is prevailing leaf characteristics: sclerophyllous = h (i.e., hard-leaved), soft-leaved = w, succulent-leaved = k, large = 1 (> 400 cm^2), small = s (< 4 cm^2) and other leaf sizes in between. A fifth subdivision is by height of the vegetation cover, for which Küchler uses eight classes, from 1 = < 0.1 m to 8 = > 35 m. A sixth and final characterization is spacing or coverage, divided into six classes. These are c = continuous (> 75 per cent) cover, i = interrupted (50–75 per cent) cover, p = parklike or in patches (25–50 per cent) cover, r = rare (6–25 per cent) cover, b = barely present or sporadic (1–6 per cent) cover and a = absent or very scarce (< 1 per cent) cover.

Küchler says that any vegetation sample can be classified by a short formula using his various criteria and subcriteria. For example, the formula B8CX refers to a very tall (> 35 m) broadleaved everygreen forest with climbers and epiphytes. Other symbols can be added for further architectural specification, for example, c for continuous cover. Küchler claims that his method works for mapping vegetation at all geographic scales. Of course, it should be kept in mind that one can outline or map vegetation realistically only in the large and intermediate map-scale ranges.

Fosberg's (1961, 1967) general purpose classification is also hierarchical, classifying vegetation cover first into three categories: closed, open and sparse. This approach is appealing from the perspective of remote sensing. A closed vegetation cover is defined as one in which the crowns of trees are mostly interlocked or the shoots of grasses are closely intermingled. An open vegetation cover is defined as one in which tree or shrub crowns are only partly interlocked and are often free-standing, so that one can see their individual outlines. The same applies to open grassland, where one can recognize individual bunches, for example. Sparse vegetation cover occurs where the spacing between individual plants is more than twice their crown or shoot outlines, or where there are groups of plants spaced widely apart. This category includes desert vegetation.

Fosberg's subdivision into closed, open and sparse vegetation cover is called primary structural grouping. Each of the three primary groups is then further subdivided into so-called formation classes. Seventeen formation classes are recognized in the closed vegetation category. These include such types as forest, scrub and dwarf scrub, which are closed woody vegetation types distinguished by stature or height. Other types of closed vegetation are closed scrub with scattered trees, open scrub with closed ground covers, tall savannah (closed tall-grass cover with scattered trees) and short-grass covers such as those found in many pastures, either natural or man-made. Also recognized among the closed formation classes are two aquatic vegetation types, floating meadows and submerged meadows. Such vegetation types are included here only if they form a dense mat.

Ten formation classes of open vegetation are recognized, such as steppe-forest, steppe-scrub and steppe (open grassland), but only four formation classes of sparse vegetation are recognized, namely, desert forest, desert scrub, desert herb vegetation and sparse submerged meadows.

Thus, Fosberg's system divides the world's vegetation cover into 31 formation classes: 17 closed, 10 open and 4 sparse.

Fosberg's next criterion (at the third level) is based on foliage persistence. The formation class of closed forests, for example, is divided into evergreen and deciduous forests, called formation groups. These are further subdivided into formation types. For example, Fosberg recognizes nine formation types within the closed evergreen formation group:

(1) Multistratal evergreen rain forest.
(2) Evergreen swamp forest.

(3) Gnarled evergreen forest.
(4) Evergreen hard-wood orthophyll forest (orthophyll = normally sized leaves).
(5) Evergreen soft-wood orthophyll forest.
(6) Evergreen broad sclerophyll forest.
(7) Evergreen narrow sclerophyll forest (narrow = needleleaved).
(8) Evergreen bamboo forest.
(9) Microphyllous evergreen forest.

The formation types are further subdivided where possible into subformation types, for which Fosberg gives particular examples. For the multistratal evergreen rain forest, for example, he cites the dipterocarp forests of Malaya and Borneo. Note that Fosberg does not mention whether the formation types are tropical, subtropical or temperate. The same applies to the schemes of Dansereau and Küchler. The architecture of forest formations is not always distinct at the borders of geographically or thermally defined belts of vegetation.

Specht's (1970, Specht *et al.*, 1974) Australian modification can be regarded as an adaptation of Fosberg's scheme to the peculiarities of Australian vegetation. Fosberg's (1967) system was accepted for classifying research sites for the International Biological Program (IBP) and similar natural, semi-natural or cultivated vegetation types. The primary IBP purpose in this context (Mueller-Dombois and Ellenberg, 1980) was to provide an inventory of important vegetation types for conservation purposes.

Australian researchers were not satisfied with Fosberg's system, since it did not take account of the special character of Australian vegetation (Specht *et al.*, 1974). On the basis of plant coverage, they recognized four instead of three structural groups, namely, dense (70–100 per cent cover), mid-dense (30–70 per cent cover), sparse (10–30 per cent cover) and very sparse (<10 per cent cover). On the basis of stature or height they recognized five classes of predominantly woody vegetation and two classes of predominantly herbaceous vegetation. The five height classes are defined as: (1) trees >30 m tall, (2) trees from 10 to 30 m tall, (3) trees from 5 to 10 m tall, (4) shrubs from 2 to 8 m tall and (5) shrubs from 0 to 2 m tall. In contrast, Fosberg (1967) outlines only three stature classes for woody plants, namely, forests (major crown biomass above 5 m), scrub (major crown biomass between 0.5 and 5 m) and dwarf scrub (crown biomass mostly below 0.5 m).

Specht's two classes of Australian herbaceous vegetation are defined as hummock grasses (from 0 to 2 m tall) and herbs (including mosses, ferns, hemicryptophytes, geophytes, therophytes, hydrophytes and halophytes). Fosberg recognizes tall grass, short grass, broadleaved herb vegetation, closed bryoid vegetation, submerged meadows and floating meadows.

The differences between Fosberg's and Specht's schemes are not fundamental. However, the Australian scheme of four coverage classes, more classes of height to distinguish between types of forest (which in Australia are mostly comprised of different species of *Eucalpytus*), and the emphasis on hummock grasses are apparently more appropriate for Australian vegetation. The Australian scheme, moreover, has a certain appeal because of its simplicity. The two major classifying criteria, cover and height (including life form), are arranged in a table in which the four cover classes are the column heads and the stature classes constitute the row heads. The table body gives the formation type names that result from the combination of stature and cover. Twenty woody formation types are given, such as tall closed forest, closed forest, low closed forest, etc. Among the herbaceous formations are two hummock grass formations (sparse and very sparse) and 19 other herbaceous formations. Oddly enough, however, savannah formations (i.e., grassland with scattered trees or shrubs) are not emphasized in Specht's scheme.

Informal synusial approaches have also been found useful for classifying and mapping vegetation architecture (Mueller-Dombois and Ellenberg, 1974). The term 'synusia' (Gams, 1918; Lippman, 1939) pertains to another structural vegetation-unit concept, namely, the species of the same life-form type growing together in the same community. In its simplest form a synusia can be described as a structural subunit within a plant community. (Daubenmire (1968) has used the term 'union' for the same concept.) That is, the tree stratum, shrub stratum, herb stratum and moss carpet on a forest floor, which may all occur together as overlapping layers in a single forest stand, can each be referred to as a generalized synusia. However, different life-form types may occur within the same layer, and these are considered to form the synusiae. For example, the tall-tree stratum of a forest may contain two tree-life forms, evergreen and deciduous trees. Each is said to form a separate synusia. Synusial approaches to classification have been reviewed by Barkman (1978). They can be used in mapping and as a means of connecting vegetation architecture and floristics with vegetation structure and function. The basis of the synusial approach is a life-form classification that includes both structure and function.

The simplest approach is the layer-diagram method. A sample stand or *relevé* can usually be described by its horizontal stratification, or layering. The limits of the horizontal strata are indicated by the various height levels of the major crown or shoot biomass. These can be arbitrarily determined, but they should follow the height stratification if it is indicated in the community. The horizontal extent of each layer can be estimated or measured.

Once the height limits (in metres) and coverage (usually in per cent) of each layer are recorded for sample plots or belt-transects, the information can be presented in a simple layer diagram. The diagram shows the height limits on the y-axis and the percentage of cover of each layer on the x-axis. Examples are given in Figure 2.2.

Such layer diagrams would appear to be useful tools for mapping plant biomass from sample plots. The quantitative distribution of forest layers indicates the vertical and horizontal distribution of the plant biomass of a stand. Accuracy could be increased by allometry, and the percentages of wood and foliage could be shown by subdivision of each layer.

Another widely known method is the profile-diagram method of Davis and Richards (1933–34) and Beard (1946, 1978). The method consists of outlining a sample belt-transect. Belt width depends on stocking density, and length may be adjusted to represent the size dimensions and within-variations of a vegetation. In forests, a length of 50 to 70 m is usually adequate for one sample. But several profile diagrams are needed for adequate sampling of the same forest type. The positions of trees above a certain size are first mapped to scale in a vertical map projection, using a string grid and metre tape. Next, the horizontal or profile view is obtained by outlining or drawing tree sizes, crown forms, crown length, branching and stem characteristics in silhouette view from a certain distance alongside the belt-transect. Photographs may help, but they are not sufficient in themselves, primarily because their backgrounds are not clear.

Two such profile diagrams are shown as Figures 2.3 and 2.4. These were prepared for the same montane rain forest as the layer diagrams shown in Figure 2.2. Profile diagrams give us certain details about major species composition, but they also portray vegetation architecture by providing considerable detail about vertical and horizontal spacing and the distribution of wood and foliar material. One can also identify on them certain life-form functions, such as the deciduous or evergreen composition of tree species. The preparation of profile diagrams requires considerably more time than that of layer diagrams, and they usually cannot be used to show undergrowth vegetation.

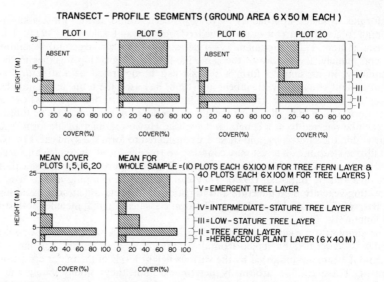

Figure 2.2   Layer diagrams of a tropical montane rain forest in Hawaii. The sample plots were taken systematically in a homogeneous *Acacia koa–Metrosideros polymorpha–Cibotium* spp. (tree fern) forest of 80 ha size. They show considerable variation in structure, particularly with regard to the presence and absence of the emergent tree layer (from Cooray, 1974). The samples relate to the Kilauea rain forest located at the N-end of Transect 2 on Figure 2.8, p. 70. The following figures (2.3, 2.4, 2.5 and 2.6) all relate to the same forest

These limitations are overcome in a third display method, the plant life-form spectrum. Such diagrams allow portrayal of the entire synusial structure of plant communities. For classifying the individuals or species of a sample stand into synusiae, one can profitably use the plant life-form classification of Raunkiaer (1937).

Raunkiaer's system is a structural-functional classification of plant types. It includes important functional aspects in the sense that plants are classified according to their seasonal behaviour. The five basic Raunkiaerian plant types are: (1) phanerophytes, i.e., perennial plants exceeding 0.5 m in potential height, (2) chamaephytes, i.e., perennial plants that do not usually rise above 0.5 m in height or that periodically throw off their shoots or branches when they extend beyond that height limit during the favourable season, (3) hemicryptophytes, i.e., perennial plants that dry up periodically to a small remnant shoot system close to the ground, although their dry shoot system may remain standing, (4) geophytes, i.e., perennial plants that reduce their entire shoot system during the unfavourable season and survive by

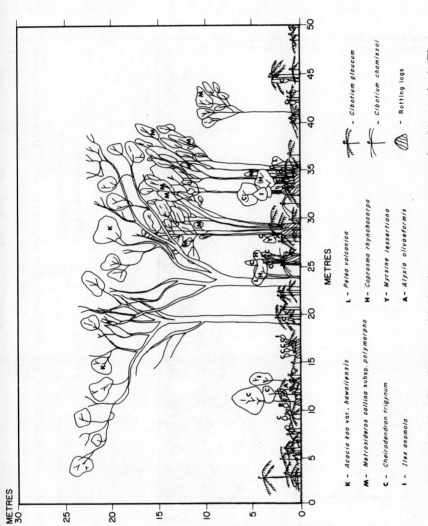

K - *Acacia koa var. hawaliensis*   L - *Pelea volcanica*

M- *Metrosideros collina subsp. polymorpha*   H- *Coprosma rhynchocarpa*

C - *Cheirodendron trignum*   Y- *Myrsine lessertiana*

I - *Ilex anomola*   A- *Alyxia olivaeformis*

— *Cibotium glaucum*

— *Cibotium chamissoi*

— *Rotting logs*

Figure 2.3 Profile diagram with dominant canopy tree species (*Acacia koa*). The diagram is a sample taken in the same forest as the layer diagrams on Figure 2.2 (from Cooray, 1974)

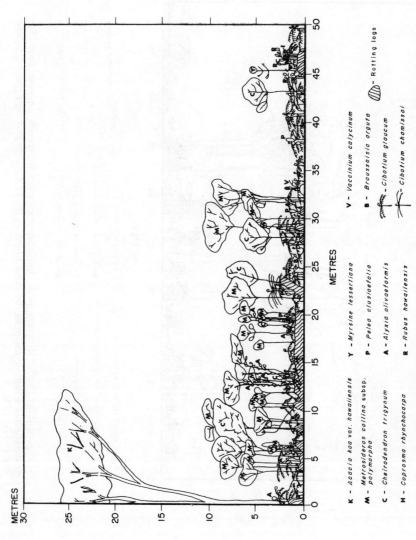

Figure 2.4 Profile diagram with 'gap phase' occupied by advanced growth of *Metrosideros collina*. Diagram represents another sample taken in the same forest as the layer diagrams on Figure 2.2 (from Cooray, 1974)

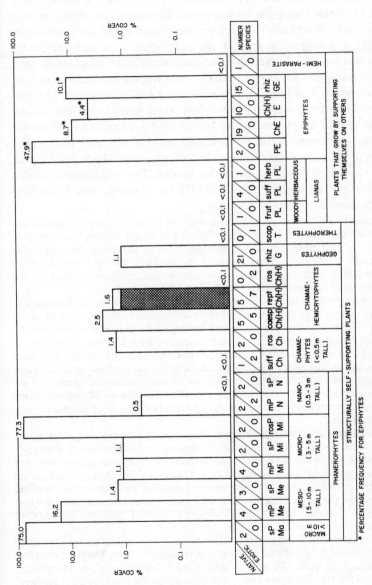

Figure 2.5 Quantitative life-form spectrum of vascular plants in a tropical montane rain forest in Hawaii. Refers to same forest as Figure 2.2. ▨ = cover of exotic species >0.1 per cent. Symbols: s = sclerophyllous; m = malacophyllous (soft-leaved); ros = leaves or fronds in rosettes; suff = suffrutescent (semi-woody); caesp = caespitose (branched from base or bunch-habit); rept = reptate (creeping or matted); rhiz = rhizomatous (modified stem imbedded in soil or organic matter); scap = scapose (single stemmed); frut = frutescent (woody); herb = herbaceous (from Mueller-Dombois *et al.*, 1981)

underground storage organs and (5) therophytes, i.e., annuals which survive only in seed form during the unfavourable season.

These five plant life-forms are the structurally self-supporting types. There are also plant life-forms that grow by supporting themselves on others, such as epiphytes and lianas. Raunkiaer's system was developed in the temperate zone, and his system was originally seen as an adaptation to a summer–winter seasonality. It was found, however, that the same basic plant adaptations have also evolved in response to seasonal rainfall in the tropics and subtropics. The plant life-form system has been further modified by Ellenberg and Mueller-Dombois (1967b) to emphasize structural-functional plant behaviour during the favourable season and to make the system also applicable to the humid tropics. Further refinement is still possible and perhaps desirable with regard to such different functional traits as shade tolerance and shade intolerance, and growth rates when these become better known. The architectural studies of tropical trees by Tomlinson and Gill (1973), for example, may be usefully combined with the existing life-form system.

It has also been found that the functional aspects of life-form types are related not only to seasonality but also (in different proportions and detail) to such characteristics as resistance to fire and other mechanical damages, such as herbivory. Moreover, the member species of a narrowly defined synusia are usually strong competitors for similar resources in a community ranking in competitive relationship next to the individuals of the same species (Mueller-Dombois and Ellenberg, 1974).

The diagrammatic representation of such life-form spectra or synusial structures is done in the form of simple histograms, where the x-axis serves to present each life-form type or synusia side by side and the y-axis and histogram blocks serve to indicate the quantity of each synusia. Figure 2.5 is such a quantitative life-form spectrum or synusial diagram and was prepared for the same Hawaiian rain forest as the layer and profile diagrams shown in Figures 2.2, 2.3 and 2.4. Of these, the life-form spectrum gives the most complete information. With this diagram comes a complete species list on which each species is identified by its life-form type. For the purpose of mapping plant biomass, however, the simpler profile diagrams—particularly the layer diagram—may be the most appropriate.

It may be noted that all of the architectural classification schemes discussed earlier make use of synusial structure in varying degrees of detail and generalization. Fosberg's system, for example, could easily be portrayed in the form of layer diagrams, with one layer diagram accompanying each formation type. Dansereau's stylized profile-diagram method is closely related to Beard's more informal method, but Dansereau's method gives more detail on the lesser synusiae. In information content, Dansereau's method stands somewhere between Beard's profile-diagram method and the quantitative synusial diagram.

These diagrammatic data display methods can be used as mapping tools in the sense that they help to define the structural patterns of a vegetation cover. They can therefore be used to define map units. As an example, Figure 2.6 shows a very large-scale vegetation map of the Hawaiian rain forest plot from which the previous diagrams were prepared. Four patterns were recognized by using these methods, and the patterns were then mapped from a very large-scale colour aerial photograph (1:1500; i.e., 1 cm = 15 m). The map shows structural-dynamic patterns, i.e., patterns that are not related to any underlying pattern of site or habitat variation.

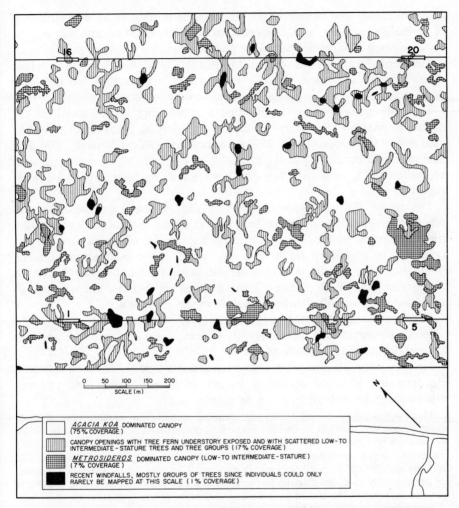

Figure 2.6 Vegetation map of montane rain forest showing four structural-dynamic patterns at a very large scale (1:8333). An 80 ha research site inside a larger homogeneous montane rain forest (the Kilauea rain forest) on the east flank of Mauna Loa. Site is located at the N-end of Transect 2 on Figure 2.8 (p. 70) (from Mueller-Dombois *et al.*, 1981)

2.4.2.2 Vegetation Floristics

Classifications based on floristics use existing patterns of plant species distribution as the primary means of mapping. There are two approaches. The first uses species distributions to define floristic provinces, while the other uses species distributions to define intrinsic species patterns (i.e., the noda or phytocoenoses). The first approach, although important to an understanding of vegetation patterns, is not relevant to the CO_2 question and will not be discussed here. The species pattern or noda approach is discussed briefly.

The term 'nodum' was coined by Poore (1955) for denoting a plant community of any rank. Poore's term was a replacement for the term phytocoenosis.

The only contemporary classification and mapping system based on the nodum approach is the Braun–Blanquet system (1928, 1965), which is widely used in central Europe (Tüxen, 1970; Westhoff and van der Maarel, 1978). The field and data-processing methods are based on a wealth of experience with vegetation and have proven their usefulness in both the temperate and tropical regions.

The major floristic approaches which can be used for classfying and mapping plant communities are based on sampling the floristic composition through sample stands, or relevés. Samples can be chosen systematically, randomly or selectively. In the last case, which is the most common, aerial photographs are usually used in combination with ground reconnaissance to distinguish preliminary patterns. These patterns are then documented in deail or subsequently modified through analysis of sample stands.

Each sample stand should be large enough to contain a cross-section of the species representative of the community. The basic data recorded are a species list and the quantities of each species.

Once an adequate number of vegetation samples (relevés) has been collected, the data are analysed according to one of several techniques: two-way tabulation, polar ordination and dendrogram cluster.

These approaches are all based on sampling the floristic composition of given vegetation covers through a number of sample stands or relevés. Samples can be placed systematically, randomly or selectively. In the latter case, which is the most commonly practiced, aerial photographs are usually used in combination with ground reconnaissance to distinguish preliminary patterns. These are then documented in detail or subsequently modified through the analysis of a number of samples. The number of samples varies with the size of the area and variability of the vegetation cover.

Each sample stand should be of a certain minimum size to contain a species composition nearly representative of the community or vegetation segment sampled. The basic data recorded in each sample is a species list (as complete as possible) and the species quantities.

Once an adequate number of vegetation samples (relevés) has been collected, the data are analyzed and synthesized according to either one of the above-mentioned three techniques or their derivatives.

The two-way tabulation technique. This is the original data-processing or synthesis tabulation technique of Braun-Blanquet, which has in recent years been variously simulated through computer application (Cêska and Roemer, 1971; Lieth and Moore, 1971; Spatz and Siegmund, 1973; Mueller-Dombois and Ellenberg, 1974; Westhoff and van der Maarel, 1978). It consists of entering the sample stand or relevé information together into one initial table, the so-called 'raw table.' Relevé numbers are entered in any order in the table head, one column for each relevé. Species names are entered on the left side of the table also in any order, one species in each successive row. Species quantities are entered into the table body for each species and the relevé in which the species was recorded.

After all species and relevés are entered, the raw table is then reordered into a 'constancy table'. This process involves counting in the raw table the number of times each species is present in the stands and recording this value (called constancy or frequency) on the right side of each species row. When this is done, species with high constancy values are temporarily eliminated from further consideration. The same is done with species of low constancy values.

The reason for taking these two steps is that species present in all or nearly all of the relevés cannot be used very well for subdividing the relevé set, since they are characteristic of the whole set. Similarly, species present in only one or two or a small proportion of the relevés cannot be used for subdividing the set because their presence is too sporadic to indicate trends.

The next step in the process involves only species with intermediate constancy values. Certain arbitrary threshold values are usually used to define the intermediate range of constancy, such as from 3 per cent to 65 per cent. These ranges are adjustable to the data set, however. The constancy table is then prepared by noting all species of intermediate constancy from high to low.

The third step involves scanning the table for species that occur together in a number of relevés. In other words, the analyst looks for species which have a high degree of association in their distribution. If he does not find such associated species, he may find single species which show a distribution trend which differs more or less strongly from that of others.

There may be a second set of species which shows a tendency to have a distribution trend which is more or less exclusive to the first set or single species. This second set is then also identified. It is usually possible to isolate at least two sets of mutually more or less exclusive groups of species which show a greater within-group similarity of distribution, i.e., some degree of within-association. For this purpose it is useful to set threshold values. For example, one might look for two species which occur together in at least 50 per cent of their relevés but occur alone in no more than 10 per cent of the other relevés. Other threshold values sometimes used are 66/10, 66/20 or 50/20. Cêska and Roemer (1971) offer five options.

Once two more or less mutually exclusive distribution trends are determined, the species involved are transferred to a third table, called an 'ordinated partial table'. The associated species of the first group are listed one under the other, and the second mutually exclusive group (or individual species, as the case may be) is listed beneath the first group. An order will usually become self-evident at this point. Further improvement may be obtained by consolidating the distribution trends through rearranging the relevé order, so that one associated group appears on the left side of the table and the other on the right. Of course, there may be more than one or two associated species groups, particularly in larger sets of relevés. The next table is obtained by rearranging the species and relevé order until the distribution trends of the associated and differentiating species groups or individuals are most evident.

When this is done, all of the hitherto excluded species are reentered in sequence, usually following the species used to differentiate the set of relevés. This final table is called a 'differentiated table'. It shows no differences in data from the raw table, but the species and relevé order has been rearranged so that the clearest species distribution and association trends are now evident.

The ecological trends associated with the species and relevé patterns are then searched for. The underlying ecological gradient may be related to soil moisture, topographically related factors, successional or dynamic relationships, or a combination of influences. Correlations with certain factors or combinations of factors may be established and may serve to suggest causal relationships or hypotheses. The next step may involve experimental analysis of the distribution hypotheses or the mapping of the patterns through the use of the differential or key species which have been worked out in the differentiated table. Some of these species may be dominants (in terms of height or cover). Mapping is usually done through a combination of ground observation and remote sensing.

Thus, the two-way tabulation technique provides, among other things, for determining key species for the mapping of floristically defined vegetation types.

The polar ordination technique was introduced by Bray and Curtis (1957) as an alternative data display technique. They felt that a geometric display of individual relevés in relation to one another according to their own floristic content would show more adequately the varying nature or continuum trends of the vegetation.

This method consists of first calculating the similarity relationships between each and every pair of sample stands or relevés in the set under comparison. The formulae used for calculating floristic similarities between stands are generally based on the proportion of common species in two relevés or on the quantitative proportion of common species in each comparison. Most commonly used is the quantitative modification of Sørensen's (1948) index of similarity proposed by Motyka *et al.* (1950). However, there are many other indexes which are equally useful. Many papers discuss the various merits of the indexes, but the usefulness of any index depends primarily on the vegetation itself (Mueller-Dombois and Ellenberg, 1974). Once the similarity indexes for the relevé set are determined, they are listed in a similarity matrix.

Polar ordination is the next step. This involves a geometric display of the similarity matrix in as accurate a constellation as possible. Because of the multivariate relationships of the similarity values, this is by no means easy. In the original Bray and Curtis version, the two stands or relevés with the greatest dissimilarity were selected to form the end points of the *x*-axis. These stands are then plotted, using their dissimilarity as the measure of geometric distance on the axis. All of the other relevés are then plotted on the same axis by using for each relevé the two geometric distance measures (or dissimilarity values) that each shares with the relevé at each end of the axis. The plotting procedure was originally done by graphical procedures. Later, the Pythagorean theorem introduced for this step by Beals (1960) was used.

When all of the relevés are plotted, one will usually find that their geometric distances do not display the dissimilarity values of the matrix very well. In other words, the geometric distances show a great deal of distortion along the first axis. For this reason, a second axis (the *y*-axis) is established by using a relevé near the centre of the *x*-axis which displays a high dissimilarity to the two reference relevés at the end points of the *x*-axis. This 'poorly fitted' relevé then becomes the first end point on the *y*-axis. The relevé most dissimilar to the first becomes the second end point. All of the remaining relevés are once again plotted with regard to the dissimilarities they share with this second pair of relevés. Thus, each relevé is positioned on each axis.

At this point, usually, the ordination is tested by comparing the geometric distances of the two-axis plot with the dissimilarity values shown in the matrix. If there is a significant correlation between the two, the ordination is considered an acceptable display of the values of the similarity matrix. If no significant correlation appears, one can either try different relevés as end points of the axes or establish a third axis, a *z*-axis. The *z*-axis can then be used to develop a three-dimensional display.

Much research has gone into finding a better way of selecting the end-point relevés for the *x*- and *y*-axes. This work has been reviewed by Whittaker and Gauch (1978), McIntosh (1978) and Cottam *et al.* (1978). The major problem is to find end-point relevés which are strongly dissimilar but which at the same time share some resemblance to the rest of the relevés to be ordinated. Extreme end-point relevés may simply be sample stands that have little in common with the rest of the set.

A large number of other ordination methods have been proposed—e.g., factor analysis (Dagnelie, 1960, 1978) and principal component analysis and position vectors technique (Orloci, 1966, 1967, 1978), but the modified polar ordination technique developed by the Wisconsin School remains one of the most powerful and robust for this approach (Whittaker and Gauch, 1978).

One value of ordination is in finding underlying environmental factor gradients or successional gradients that can be used to explain the floristic vegetation patterns displayed on the diagrams. Its second value is to test intuitive classifications of plant communities by mathematical procedures. Polar ordination, therefore, is useful for plant community classification, but it was not really designed to be a classification technique.

The dendrogram technique, sometimes called the cluster analysis technique, is a mathematical tool explicitly designed for classification purposes, including plant community classification (Sokal and Sneath, 1963). Basically, the dendrogram technique also involves a diagrammatic display of the content of a similarity matrix and in this respect is similar to polar ordination. But instead of displaying relevé similarities or dissimilarities by geometric distance, dendrograms display them in the form of 'clusters'. These clusters are mathematically computed levels of similarity, indicated on the y-axis of each dendrogram, at which two or more relevés are joined. The joining of certain relevés at certain levels of similarity indicates a cluster, or community type.

The centroid, polythetic and agglomerative dendrogram technique begins with a search for the two relevés with the greatest similarity. These two are then compared with the relevé that shares the next highest similarity to one of the first two relevés; or the first two relevés are averaged into a single synthetic relevé. In that case a new set of similarity indexes is computed between the synthetic relevé and the others. Although the details of the various procedures differ, the underlying goal is to sequentially compare all relevés to one another. The relevés are clustered in sequential algorithms at continuously decreasing levels of similarity until they all are clustered.

It is quite rare for relevés to display complete similarity. But at a very low level of similarity, say five per cent, all of the relevés may be united in a single cluster. The reason is that they are likely to have at least one species in common.

The ecologically most important information in any dendrogram is found not in the very high-similarity or very low-similarity clusters but in those occupying the intermediate range. In this intermediate range one usually finds the clusters that provide usefully generalized information.

The problem in the dendrogram technique is to identify the set of intermediate-range clusters which are useful for community classification. (The same problem exists in the synthesis tabulation technique when the investigator looks for associated species groups in the constancy table.) Arbitrary guidelines have been devised (Orloci, 1978) for objective cluster recognition, but the problem has not yet been totally resolved. It is difficult to make plant community classification an automatic matter. The reason for this difficulty is that similarity levels within or between ecologically meaningful clusters or plant community types cannot be reduced to one level alone. If it was attempted, the natural variation in vegetation would be violated, and the result would be an arbitrary or meaningless classification.

One further point should be mentioned. The two-way tabulation technique is superior in two ways to the polar ordination and dendrogram techniques. The differentiated table displays the species by name and with their quantitative variation among the whole set of relevés. The other two techniques are more abstract, in that the investigator must make a special determination of the species which are responsible for the displayed variation.

Of course, one can usually also use these two techniques for arranging and displaying species relationships. One can, for example, ordinate by species rather than by relevés, or one can build a dendrogram on species distribution similarities rather than relevé similarities. But this is not the same as building species–relevé blocks and seeing the species distributions at the same time on the display tool as in a differentiated synthesis table. An ordination tool called 'reciprocal averaging' introduced by Hill (1973) is said to display species and relevé groupings in coordination (Whittaker and Gauch, 1978), and this may become a tool as powerful as the two-way tabulation technique.

Another powerful technique in classifying vegetation from the standpoint of floristics is the structural analysis of species populations. The tree species-population analysis technique is essentially a forest mensurational technique (Meyer, 1952) which has proved to be of equivalent ecological value to the other techniques discussed.

This technique involves enumerating species individuals by size in sample quadrats (belt-transects, circular plots or one of the more popular distance measures). Usually used as a measure of size are diameter at breast height (dbh) for trees over two metres tall and height for seedlings and saplings under two metres tall. Both measures can also be applied at the same time, or others can be used, such as basal area and volume. The number of trees enumerated in a sample should be at least 30 to 50 in each class (i.e., trees over or under two metres) so that definite trends can be tabulated or plotted into curves (Mueller-Dombois and Ellenberg, 1974). The curves can then be interpreted. For example, an inverse J-shaped distribution would indicate a large number of small juvenile trees and perhaps normal mortality trends in successively larger groups, indicating a stable population. Such a trend may be indicative of a climax tree species. A large number of large-sized trees with few or no small trees would indicate a poorly reproducing population, possibly a pioneer tree species. Considerable work has gone into interpreting curve trends mathematically and ecologically (Leak, 1965; Hett and Loucks, 1971; Goff and West, 1975), but they are only trends and do not convey definite answers. Additional information has to come from autecological and synecological studies of the communities and environment of the study area. In this respect the number/size distribution curves used for the analysis of the population structures of tree species are like the display techniques discussed earlier. They provide trend indications rather than conclusive answers, but they are still extremely valuable.

It should be noted here that one problem in the study of tropical rain forest ecology is the practical impossibility of determining tree ages. But age estimates are useful additional information in conjunction with number/size distribution curves.

The techniques that have been discussed here are useful for classifying forest communities into successional groups, provided that spatial variations in habitat are understood and taken into consideration. The extrapolation of successional trends from data on trees and stands occurring in different locations is scientifically suspect unless it can be shown that the habitats are similar or identical. When this is demonstrable, data on population trends add a powerful tool for determining and classifying dynamic relationships among plant communities. Hence, a thorough classification of forest communities should make use of both spatial and structural-temporal analysis techniques.

2.4.3 Problems and Measures in Mapping Tropical Vegetation

Tropical vegetation has been studied much less than temperate zone vegetation, although the taxonomic diversity of tropical rain forests is a well-established fact. The high diversity makes the definition of tropical plant communities and their mapping particularly difficult. However, there are several simplifying approaches.

2.4.3.1 Bioclimatic Methods

Except perhaps for Thornthwaite's method, the bioclimatic classification methods were formulated with tropical vegetation in mind. This is apparent in Köppen's classification, clearly, in the Holdridge system and also in Walter's climate-diagram method. Whitmore (1975) found the simple seasonality scheme of Mohr (1933) to be most appropriate for matching climate with the major zonal forest types in southeast Asia. Mohr considered monthly rainfall in Indonesia to be 'wet' when it exceeds 100 mm and to be 'dry' when it measures less than 60 mm. On this basis he developed a formula, the number of dry months divided by the number of wet months. Whitmore used the climate diagrams of Walter and Lieth (1960) to extract seasonality data for a nomogram which shows the mean number of dry months plotted over the mean number of wet months. A diagonal axis drawn from the value of 12 dry months on the y-axis to the value of 12 wet months on the x-axis shows the range of Mohr's quotient (Q), from infinity to zero. On this nomogram Whitmore defines four bioclimatic zones for southeast Asia:

Type A = Wet or perhumid tropical zone, that is, up to 1.5 dry months ($Q = 0.143$)

Type B = Moist or humid tropical zone, that is, 1.5 to 3 dry months ($Q = 0.143$ to 0.333)

Type C/D = Moderately dry tropical zone, that is, 3 to 6 dry months ($Q = 0.333$ to 1.0)

Type E/F = Dry tropical zone, that is, climates with more than 6 dry months ($Q = 1.0$ upwards)

On Whitmore's small-scale bioclimatic map, much of the Malay Peninsula, Borneo, Sumatra and New Guinea are shown as Type A. The Type B climate is shown for southeast Java, southern and eastern Borneo and the central Philippines. On western Java, southern New Guinea, Thailand and the northern and southern Philippines one finds a moderately dry (Type C/D) climate. On the islands east of Java and in much of northern Australia one finds a dry tropical (Type E/F) climate. The tropical rain forest of northern Queensland is shown as occurring in a moderately dry (Type C/D) climate.

This dry month/wet month ratio can, of course, only be used for tropical bioclimatic mapping. Rainfall seasonality is apparently the most important environmental characteristic related to spatial differentiation in tropical vegetation types. Only Walter's climate-diagram method can readily be used for detailed, region-specific, bioclimatic mapping. It should also be pointed out that Mohr's system can only be used to characterize tropical lowland climates. Tropical mountain climates require special consideration of temperature relationships. These can be obtained from Walter's climate diagrams.

Kartawinata (1977) used Mohr's bioclimatic system in combination with Walter's diagram method in somewhat modified form to characterize 24 ecological zones in

Indonesia. He also applied Krajina's (1973) biogeoclimatic zonation scheme as a systematic tabulation device.

2.4.3.2 Landscape Classifications

The landscape classifications discussed earlier are all applicable to small- and intermediate-scale mapping in the tropics. Krajina's bioclimatic zonation scheme, developed primarily for British Columbia, has been applied to the Hawaiian Islands (Krajina, 1963) and to Indonesia (Kartawinata, 1977). Gaussen's system is widely applied, particularly in the tropics of southeast Asia. Moreover, his system is closely related to the UNESCO system, and Ellenberg's world ecosystem classification builds on the UNESCO system.

Beard (1978) reviewed the physiognomic approach to classifidation with particular attention to tropical vegetation. He emphasizes the confusion among Anglo-American ecologists (also reviewed in detail by Shimwell, 1971) which was caused by the successional classification scheme of Clements. Clements (1916, 1928) considered a bioclimatic unit or a vegetation zone to represent a formation in spite of the various physiognomic units, such as grassland, scrub and forest, which are usually encountered in a vegetation zone. Clements regarded these variations as successional types which would all converge in time toward the climatic climax. Moreover, Clements did not define these vegetation variations in a zone physiognomically but merely floristically by dominant species.

Tansley and Chipp (1926) argued against Clements' idea that all vegetation variations in a climatic zone would converge to a climatic climax and recognized other stable types as edaphic or topographic climaxes. Because of this argument, physiognomic or more detailed structural and floristic studies were neglected in the English-speaking countries but not in central Europe or Russia.

British ecologists who worked in the tropical countries of the commonwealth then re-emphasized the need for physiognomic classification and mapping. This began with Burtt-Davy (1938) and Champion (1936) and continued through Richards (1952) in particular, who together with Davis devised the profile-diagram method (Davis and Richards, 1933–34) as a realistic tool for describing vegetation.

A convergence of the purely physiognomic approach with the landscape approach is apparent in Beard's (1955, 1978) formation-series approach. Beard uses profile diagrams in a more generalized context by showing typical formations (based on vegetation architecture) in relation to climatic gradients. This is a very effective approach which should be used wherever possible as a tool for map interpretation. It has been applied by many authors, including Whittaker (1975) in his discussion of major ecoclines. But it should be understood that both Beard's formation series and Whittaker's major ecoclines are landscape units rather than real vegetation units. They demonstrate potential physiognomic vegetation, not existing vegetation.

A more detailed approach is Whitmore's (1975) treatment and classification of the tropical rain forests of southeast Asia. Whitmore recognizes 16 types of tropical forest formations:

(1) Tropical lowland evergreen rain forest.
(2) Tropical lower montane rain forest.
(3) Tropical upper montane rain forest.
(4) Tropical subalpine forest.
(5) Heath forest.
(6) Forest over limestone.
(7) Forest over ultrabasic rock.
(8) Beach vegetation.
(9) Mangrove forest.
(10) Brackish-water forest.
(11) Peat swamp forest.
(12) Fresh-water swamp forest.
(13) Seasonal swamp forest.
(14) Tropical semi-evergreen rain forest.
(15) Tropical moist deciduous forest.
(16) Other formations of increasingly dry seasonal climates.

The last two formation types are described as monsoon forests, while the first 13 are tropical rain forest types. Poore (1978) considered Whitmore's 16 formations more useful for the classification and conservation purposes of the International Union for the Conservation of Nature (IUCN) than the 14 types suggested in the UNESCO scheme. Poore believes that Whitmore's types can also be extrapolated quite satisfactorily to the American and African tropics.

The difference between Whitmore's types and Whittaker's major ecoclines is that Whitmore recognizes not only the relationships of vegetation to bioclimates but also to major variations in soil and topography. Variations in relation to topography are also recognized in the UNESCO system, but Whitmore gives more detail on edaphic differentiation. Rodin *et al.* (1975) used a similar landscape formation concept for their productivity and phytomass estimates. They called their units soil-vegetation formations and recognized 10 such units for the humid tropical belt.

Whitmore's 16 formation types can be grouped by environmental factors as follows:

(a) Formations dominantly controlled by lowland climates = 4 (1, 14, 15, 16).
(b) Formations dominantly controlled by topographic factors, mostly mountain climates and associated factors = 3 (2, 3, 4).
(c) Formations dominantly controlled by edaphic factors = 9 (5, 6, 7, 8, 9, 10, 11, 12, 13).

Whitmore's formations are not physiognomic formation types; they are landscape units or landscape formation types. Very little emphasis is given to vegetation architecture in Whitmore's types. In fact, one may expect to find several physiognomically or architecturally different vegetation types in each of Whitmore's units. This becomes further apparent from Whitmore's observation that most of the tropical rain forest in southeast Asia has been disturbed by shifting cultivation and thus is not original but secondary forest. Such secondary forests may differ from place to place even on the same soil substrate, and one may expect to find a mosaic of successional vegetation types within each of Whitmore's formation types. These secondary forests and successional types will have different physiognomies or architectural characteristics and thus different biomass relationships.

2.4.3.3 Methods for Mapping Existing Vegetation Structure

The profile-diagram method developed by Davis and Richards (1933–34), the more formalized profile-diagram procedure of Dansereau (1951), the schemes of Küchler (1967) and of Fosberg (1967), and the layer-diagram and life-form spectrum methods all are useful for mapping tropical as well as temperate forest architecture.

Küchler and Sawyer (1967), for example, mapped an area covered with tropical moonsoon forest vegetation in northwest Thailand at a scale of 1:30 000 using Küchler's system. The authors defined phytocoenoses by both architectural criteria and characteristic species composition. Fosberg's system has been used to map two national parks in Sri Lanka, one covered dominantly by evergreen sclerophyll monsoon forest, the other dominantly by deciduous scrub vegetation (Mueller-Dombois, 1969, 1970). The maps were prepared from air-photo mosaics at a scale of 1:32 000. Mapping by dominant species was not possible in this vegetation because of the great species diversity in both woody and herbaceous vegetation types. Approximately 20 to 30 purely structurally defined types were mapped for each national park (each was approximately 80 000 to 100 000 ha in size). Where necessary, the Fosberg system was expanded to recognize finer subdivisions. An aerial photo key for objective mapping procedures was developed (Mueller-Dombois and Ellenberg, 1974: 489) which provided for quantitative definitions of map units, so that a trained interpreter could carry out independent mapping at the same level of detail and unit integration.

In the tropical rain forest territory of Hawaii forest types have been mapped at a scale of 1:48 000 by using both dominant species and non-floristic structural criteria (Mueller-Dombois *et al.*, 1977). The map units were systematically worked out, again following Fosberg's mapping principles, to obtain internationally comparative information values. In fact, it would be insufficient to map vegetation in the Hawaiian rain forest by dominant species because of the broad ecological amplitude of *Metrosideros polymorpha* and the tree fern species of *Cibotium*. The problem is similar to that in central Europe, where undergrowth vegetation and a combination of characteristic species provide better large-scale mapping criteria than dominant species do.

The usefulness of the structural (non-floristic) classification approach has already been tested in one case. Becker (1976) applied morphological data collected from tree ferns (*Cibotium* spp.) to test an existing map classification (Mueller-Dombois and Fosberg, 1974) of the montane rain forest in and near Hawaii Volcanoes National Park. By applying tree fern data, such as stem density, height, top-trunk diameter and number of fronds to a multivariate analysis using the dendrogram technique, Becker found that these tree fern data alone were sufficient to arrange his sample stands into an ecological series. This meant that tree fern architecture could be used as an indicator of habitat in that area.

Webb (1968, 1978) has pioneered physiognomic–architectural approaches for the classification of Australian rain forests. He and his collaborators (Webb *et al.*, 1970, 1976) have dealt with such data by numerical methods, particularly by using the two-way tabulation and dendrogram techniques. Webb found a good correlation between purely structural (non-floristic) characteristics and environmental factors. His dendrogram classification of the northeastern Australian tropical rain forest resulted in the nine types given in Table 2.4 (synthesized from Webb, 1968, 1978).

It is interesting that all rain forests in Australia (11° to 20 °S latitude) are called vine forests by Webb (Table 2.4). This designation also applies to Webb's subtropical forests (20°–32 °S latitude). The first physiognomic criterion used by Webb is evergreenness versus deciduousness (as in Küchler's system). The term 'raingreen' in Table 2.4 is applied to Australian tropical forests with at least a minor proportion of deciduousness.

Table 2.4   Tropical rain forest types of NE Australia according to Webb (1968, 1978) 1978)

Structural types of tropical rain forest in Australia	Mean rainfall (mm)	Soil fertility (rating)
A. *Strictly evergreen*		
1. Mesophyll vine forest	3644 (seasonal)	Medium
2. Mesophyll fan-palm vine forest	3644 (seasonal)	Medium (seasonal swamp)
3. Notophyll vine forest	1594 (monsoonal)	Calcareous or siliceous sands
B. *Raingreen*		
4. Complex mesophyll vine forest	3644 (seasonal)	High
5. Mesophyll feather-palm vine forest	3644 (seasonal)	High–medium (seasonal swamp)
6. Mixed mesophyll-notophyll vine forest	3644 (seasonal)	Medium
7. Semi-deciduous mesophyll vine forest	3049 (strongly seasonal)	High–medium (alluvial)
8. Semi-deciduous notophyll vine forest	1600 (monsoonal)	Medium
9. Deciduous vine thicket	1222	High–medium–low

Below this first level, prevailing leaf size is used—i.e., mesophyll > 12.5 cm, notophyll 12.5 to 7.5 cm and microphyll < 7.5 cm. Prevailing microphyllous trees, such as those associated with the southeast Asian heath forests of Whitmore (1975), are not apparent in the Australian tropical rain forests. The term 'complex' in Table 2.4 refers to forests highly variable in leaf size, while 'mixed' refers to a prevailing mixture of two leaf sizes. In addition to leaf size, specialized tree forms (tree ferns, palms, etc.) are also used for subdividing. Spacing, or coverage, is not used in Webb's system, although it is the first-level criterion in Specht's (1970) Australian scheme and is also used in Fosberg's system and others.

Webb defines his structural types in such a way that they are useful as site indicators in a broad sense. This is apparent in that each type is characterized by either a particular rainfall regime, by type of soil, or by a combination of climate and soil. In his 1968 paper Webb provides a field key for the identification of general habitat types through the use of vegetation architecture. In this respect Webb's structural vegetation classification is primarily a landscape scheme.

Habitat identification is not implied in Dansereau's, Küchler's, Fosberg's or Specht's schemes. Those schemes are based strictly on vegetation architecture alone. An ecological meaning can be applied to such strictly architectural units, however, through the analysis of environmental, historical or dynamic-successional factors (see, e.g., Brünig, 1970).

It may be noted that schemes using only vegetation architecture are more useful for the mapping of plant biomass than those using vegetation indicators for habitat identification. That Webb's scheme belongs to the latter can be seen from the absence of coverage and stature among his criteria. These two structural characteristics are the most important ones for phytomass mapping.

However, primary production cannot be mapped by any purely vegetation–architectural scheme, since environmental factors and vegetation potential are the critical factors. Webb's scheme thus might be adaptable to the mapping of primary production in the tropical rain forests of Australia. This approach could also be extended to other tropical regions.

2.4.3.4 *Environmental Gradient Analysis*

Whittaker (1967) distinguished two approaches to environmental gradient analysis, direct and indirect. Direct analysis means analysing vegetation response to a known environmental gradient. Indirect analysis means analysing the environment in relation to a known vegetation gradient. These two approaches are clearly analogous to the bioclimatic and landscape methods. Bioclimatic mapping is a direct form of environmental gradient analysis that is applied in small- and intermediate-scale maps, while direct environmental gradient analysis is usually associated with local or regional field work and large-scale maps. Whittaker's (1956, 1960, Whittaker and Niering, 1965) mountain-slope analyses are typical examples, and so are Ellenberg's (1950, 1952) gradient analyses of agricultural weed communities in Europe. Landscape mapping, such as shown on Hueck and Seibert's (1:8 million) map of South America, is a typical case of indirect gradient analysis, but in the small map-scale range. An example of indirect gradient analysis in the intermediate to large-scale map range is Webb's (1968, 1978) approach to the subhumid and seasonal tropics of Australia.

There are problems in mapping tropical vegetation with respect to both approaches. The reason is high species diversity, which presents two different kinds of difficulties.

One problem is in identifying such a large number of species particularly in areas where taxonomic information is insufficient. It prompted Webb (1959, 1968) to look for vegetation–architecture alternatives for classifying tropical rain forests, and it is the same problem that traditionally has led to a general preference for physiognomic over floristic approaches to classifying and mapping tropical forest vegetation.

A more fundamental problem, however, is the character of the distribution of tropical forest species. Even in rather large sample plots, most of these species are represented by only one or two specimens (Ashton, 1964, 1965; Poore, 1964, 1968; Letouzey, 1978). Another, but much smaller, group of species are so widespread, however, that they show no response to seemingly wide variations in soil. A third group of species—those with intermediate ranges of distribution and abundance that might be expected to reflect spatial changes in soil substrate—often seem to be rare or absent in tropical forests. Knight (1975), for example, did not find any tropical forest species that responded significantly to soil variations in Panama. On the other hand, Ashton (1964) found forest stands to cluster in relation to soil and local topography in the dipterocarp forests of Brunei. By using ordination techniques he obtained a fair clustering of stands with similar floristics in similar habitats. The environmental gradient trends indirectly established by Ashton were supported further by Austin *et al.* (1972) using different multivariate techniques.

It is perhaps too early to generalize about the spatial relationships of tropical forest species and habitats on large-scale maps because there have been so few studies of this sort. Whitmore (1975) emphasizes that practically nothing is known about small-area variations in the distribution of tropical species in relation to variations in habitat. Small-area patterns in tropical rain forests have primarily been identified so far in a dynamic sense as gap phases, building phases and mature phases. A few people also refer to a breakdown phase (Leibundgut, cited in Mueller-Dombois and Ellenberg,

1974: 398; Zukrigl, Eckhardt and Nather, cited in Walter, 1976: 15). The breakdown phase in the montane rain forest of Hawaii, however, is so conspicuous (Mueller-Dombois *et al.*, 1977; Mueller-Dombois, 1980a) that the phenomenon has been called dieback, or forest decline. It is still popularly thought to be caused by a disease (Thompson, 1978).

These phases are the most obvious architectural variations and need urgent attention in different rain forest areas, since dynamic phasing patterns are a key to understanding ecosystem maintanence under natural conditions. But studies of pattern and process in tropical rain forests cannot advance until the responses of vegetation to habitat variations are also clarified. One can be too easily tempted to deduce dynamic relationships from structural variations which in reality are the result of spatially different environmental situations.

Direct gradient analysis and the use of vegetation–architecture characteristics as the response phenomena may be the way to proceed. The direct approach has not been used much in tropical forests. It will permit more precise determination of the vegetation potential of different habitats in a local forest management unit, i.e., on large-scale maps. Such studies are urgently needed for mapping primary production and other site capability and fragility information that is needed for tropical forest management.

2.5 A COMBINATION METHOD FOR APPLYING SATELLITE IMAGERY TO VEGETATION MAPPING AND FOREST LAND CLASSIFICATION

Satellite imagery has added a significant dimension to the array of remote sensing capabilities. In the context of this paper, two aspects of satellite imagery are particularly important. One is LANDSAT imagery. The other is the possibility of extending the mapping of actually existing vegetation from the large-scale to almost the small-scale map range. Satellite imagery— available in standard black-and-white format at an approximate scale of 1:500 000—allows one to recognize the outlines of actual vegetation even when reduced to the lower limit of the intermediate-scale range, a mapping scale of 1:1 million.

This is the map scale for which the physiognomic–ecological classification scheme of UNESCO was prepared. Thus, it should be possible to use 1:1 million satellite imagery to outline the 225 physiognomic–ecological units recognized in the UNESCO scheme if one has additional information, which can be combined with this approach. The scheme provides for the mapping of real vegetation–natural, semi-natural and cultivated. The internationally integrated terminology of the UNESCO classification goes a long way toward removing the ambiguousness inherent in many vegetation terms. This ambiguity, a long-standing handicap in global understanding of vegetation-type similarities (Burtt-Davy, 1938; Richards *et al.*, 1940; Vareschi, 1968; Latouzey, 1978), has thus been resolved, to some extent at least.

Except for a few successful tests in the Central American tropics (Küchler and Montoya-Maquin, 1971), however, the UNESCO system has not been applied over any significant area of the globe. Fontaine (1978) states that UNESCO and FAO are collaborating in applying the UNESCO classification to the production of vegetation maps at a scale of 1:1 million, but an important question remains: To what extent is satellite imagery intended for use in the 1:1 million mapping process? Without the use of satellite imagery, the 1:1 million maps will tend to become maps of potential natural vegetation—i.e., site or landscape maps—rather than maps of existing vegetation. Maps of potential vegetation are valuable for different purposes, of course, but satellite imagery has provided the capability to extend the mapping of real vegetation to broad overview scales. This has never been possible before and should allow for more exact inventories of phytomass.

2.5.1 Imagery in Mapping

The following specific proposals are made for applying satellite imagery to vegetation mapping and forest land classification.

2.5.1.1 Structural Elements

Remotely sensed images provide certain basic elements of what is sensed. Nine are generally recognized (Estes and Simonett, 1975):

(1) Size (i.e., scale).
(2) Shape.
(3) Shadow.
(4) Tone or colour.
(5) Texture (rough or smooth).
(6) Repetition of pattern.
(7) Site (i.e., the indirect indication of what is there, e.g., a black spruce swamp).
(8) Association (objects commonly associated with one another).
(9) Resolution (which puts a limit on what can be recognized on the image).

There are some striking analogies between these elements and the basic structural criteria used in mapping vegetation in the field. *Size* relates to map-scale, which determines to a large extent what kinds of vegetation structure or community type can be recognized. *Shape* relates to the outlining of vegetation units, which in some cases may be self-evident but in others very difficult to identify. *Tone* or colour variations are usually used to aid in determining the boundaries of plant communities, which may be recognizable by differences in albedo. Aside from topographic considerations, *texture* may be translatable into open, closed or sparse vegetation, the primary structural groups in Fosberg's (1967) scheme. At larger aerial photographic scales, texture may be translatable into grassland with scattered woody plants, closed grassland, scrub or forest, depending largely on albedo. *Shadow* can sometimes be translated into height of a forest on large-scale photographs. Shadow as a structural element is used for timber volume surveys from aerial photographs (Thorley, 1975). However, at this level of detail the use of stereopairs rather than shadow is the more common means for evaluating tree height and topographic variation. *Repetition of pattern* on remotely sensed images aids in the classification of plant communities according to structural similarities, while *site* in remote-sensing terminology refers to the recognition of landscape units, i.e., vegetation plus habitat. Such landscape units, which are rather generalized entities from the viewpoint of vegetation architecture, are commonly recognizable on satellite images. *Association*, which in remote-sensing terminology refers to repeating patterns of plant communities in spatial contact, provides an explanatory dimension which is often omitted from plant community classifications which have not been put to the test through mapping.

An important advantage of satellite imagery is the broad overview that these images provide. A person with some geographic knowledge may be able to recognize a particular area from small-scale satellite images without resorting to an atlas. On the other hand, small-scale images suffer from low resolution, which means that it is possible to recognize only limited amounts of detail.

2.5.1.2 Non-Technical Classifications

In the *Manual of Remote Sensing*, Thorley (1975) reviews the classification of forest land in the temperate United States and Canada, in temperate Europe, in the temperate southern hemisphere (particularly Australia, New Zealand and Chile), and in the tropical zone. He points out that forest areas in North America are mapped by dominance types, such as the Douglas-fir and the lodgepole pine forests. The same sort of forest classification is practised in New Zealand. In Australia, forests are classified for commercial purposes into three broad classes—productive, semi-productive and low-grade. Physiognomic formation concepts are apparently applied in the tropical zone, but there is no uniformity within the tropics nor within any of the temperate areas. Moreover, the manual gives almost no guidelines for mapping forest land from satellite imagery. The focus of attention is on conventional aerial photography, which is in the large-scale range, most commonly from 1:10 000 to 1:50 000.

According to Thorley, however, satellite imagery at the scales of 1:100 000 to 1:500 000 is used for a number of forestry purposes, including damage assessment (such as pollution damage, fireloss and insect and disease damage) and land classification. Indeed, there are many recent forestry papers which are concerned with the application of multispectral satellite imagery to forest land classification. Es (1976) applied such imagery to the classification of the tropical dry-deciduous forest of central India and reports that such imagery could be used to distinguish forested from agricultural areas. Within the forested area Es could distinguish two groups in terms of size, woody vegetation taller than 15 m and woody vegetation less than 15 m high. Tiwari (1976) was able to distinguish such units as forest, scrub and grass cover. Sicco Smit (1974) applied SLAR images to the mapping of tropical rain forest in

Colombia and was able to distinguish broad vegetation types on the basis of physiographic features. Wacharakitti (1975) reports that he could distinguish five landscape types in northern Thailand: evergreen forest, mixed deciduous forest with teakwood (*Tectonia grandis*), dry dipterocarp forest, rice fields and areas of shifting cultivation. Driscoll and Spencer (1972), using satellite imagery, concluded that only generalized vegetation types could be identified in Colorado, such as Ponderosa pine forest, herbaceous upland vegetation and hydrophyte vegetation. Jobin and Beaubien (1974) were able to recognize such types as *Picea mariana* and *Abies balsamea* forests on Anticosti Island from both black and white and colour imagery provided by ERTS-1, while Lawrence and Herzog (1975), using the same source of imagery for two areas in central Oregon, concluded that forest cover type could not be adequately described by species name alone. These examples may suffice to indicate that satellite imagery in the normally available intermediate-scale range can only be used to outline broad vegetation types.

2.5.1.3 Accessory Information at Satellite Image Scales: the Broad Structure

As an example of how to use accessory information at satellite image scales, Figure 2.7 gives a 1:1.3 million satellite image of the island of Hawaii. On this image one can recognize three clear patterns. These are ocean, land and clouds. But there are also certain obvious patterns on the land itself. For example, there are some significant topographic configurations indicating mountains. On the mountain in the southern half of the image one may recognize a large crater-like depression. However, it is not easy to determine the height of these mountains from the image. Two large patches of dark colour occur on the east side of the island. One might suspect that these were forest-covered areas, but even if we found out with the help of improved spectral bands and some enlargement or colour-change techniques that these dark patches are indeed forest-covered areas, how can we find out what kind of forest this is? Our approach is to use existing information mapped at approximately the same geographic scale. Such information may be used in any sequence or form, but the more important elements for vegetation mapping are climate, topography, soil and the vegetation itself.

2.5.1.4 Climate and Topography

Figure 2.8 is a topographic map of the island of Hawaii with 21 climate-diagrams. This map is reduced to approximately the same scale as Figure 2.7, i.e., 1:1.5 million. The additional information in Figure 2.7 indicates at once that the northern mountain on the satellite image is Mauna Kea and the southern mountain Mauna Loa.

Figure 2.8 also indicates that both mountains are of approximately the same height. The uppermost contour line for both mountains is 13 000 ft (3965 m).

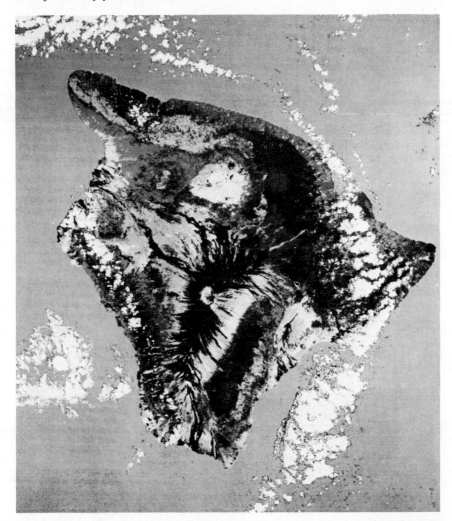

Figure 2.7 Satellite image of the Island of Hawaii at 1:1.3 million. 1973 LANDSAT image from Dec. 1978–Jan. 1979 issue of the *Plain Brown Wrapper*, an Ames Research Center publication

The exact altitude of each mountain could easily be obtained from a more detailed topographic map. The east flank of Mauna Kea slopes seaward along transect 6 from above 4000 m to sea level over a distance of only 40 km (27 mm on map).

The climate diagrams were prepared according to the method of Walter (1957). The abscissa on each diagram represents the 12 months of the year from January to January, with July in the middle (since Hawaii is in the

The role of terrestrial vegetation in the global carbon cycle

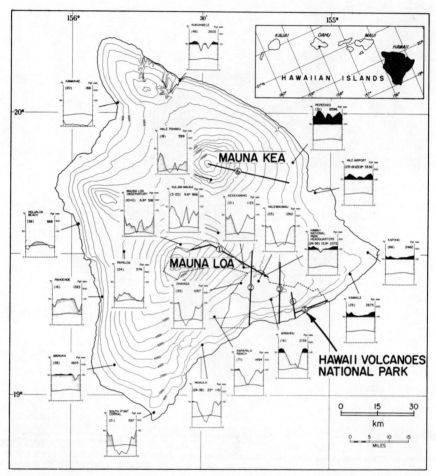

Figure 2.8   Climate-diagram map of the Island of Hawaii at 1:1.5 million (from Mueller-
Dombois, 1966a). It shows 21 climate diagrams prepared after Walter (1957). The abscissa
on each diagram portrays the months from January to January with July in the centre. The
left-side ordinate gives the air-temperature scale, from 0 upwards in 10 °C intervals. The
near-horizontal curve is the mean month-to-month temperature curve. The right-side
ordinate gives the precipitation scale, from 0 to 100 mm upwards in 20 mm intervals and
from 100 mm (the black areas on some diagrams) in 200 mm intervals. The mean month-
to-month precipitation curve is the uneven line on each diagram. Pronounced dry seasons
are indicated in black. Station name, years of record and mean annual precipitation are
shown at the top of each diagram. Transects 1 to 6 relate to ecosystem profile diagrams;
one of these (Transect 6) is shown on p. 76.

northern hemisphere).* The left-side ordinate on each diagram gives the temperature scale in 10 °C intervals, starting from 0 °C. The right-side ordinate on each diagram gives the precipitation scale in 20 mm intervals, from 0 to 100 mm. Above 100 mm the scale is reduced by 10% so that the next interval above 100 mm represents 200 mm. The black area on the diagrams indicates monthly precipitation in excess of 100 mm. Diagrams with black areas across the top are typical for rain forest areas in the tropics. The mean monthly rainfall or precipitation curve, which is plotted with reference to the right-side ordinate, is the uneven curve. The mean monthly temperature curve, plotted with reference to the left-side ordinate, is always a smooth curve that follows almost a straight line trend on each diagram. This temperature curve shows that we are in the tropics, since the difference between summer and winter mean air temperatures is very small. This curve does not change in shape because of rising altitude; it is merely displaced downward, so that at the Mauna Loa Observatory, for example, the mean monthly temperature never exceeds 10 °C, and the annual mean temperature as shown on that diagram is only 6.9 °C. This temperature curve and mean value are typical for tropical alpine environments. At the south end of the island are tropical seasonal climates, which are indicated wherever the rainfall curve falls below the temperature curve. The west coast has a reversal of seasonality into humid summer and dry winter. On the northwest side of the island is a typical climatic desert (at Kawaihae). Here the rainfall curve remains below the temperature curve throughout the year. This Hawaiian climate compares in dryness with the western Sahara Desert (Walter *et al.*, 1975).

In Figure 2.7 the dark pattern around the top cannot be forest, since this is an alpine environment. The large dark patch of the east flank of Mauna Kea, however, is a forest-covered area. The climate diagrams of Pepeekeo and Hilo Airport in Figure 2.8 inform us that this must be tropical rain forest because both diagrams show rainfall in excess of 100 mm for each month of the year. Since the forested area is topographically delimited near 2000 ft (610 mm), it cannot be lowland tropical rain forest. Its upper extent, according to the satellite image, goes to about 9000 ft (2745 m), and we may wonder whether this is all included in the montane rain forest. The climate-diagram density is not quite sufficient to give us this answer; the answer may be obtainable from other sources.

2.5.1.5 Vegetation and Soil

The vegetation zone map of Ripperton and Hosaka (1942) shown in Figure 2.1 gives us an answer to the question of where to draw the upper boundary of the montane rain forest. According to this map, the montane rain forest (zone 6)

---

* Similar diagrams for areas in the southern hemisphere show January in the centre of the *x*-axis (Walter *et al.*, 1975).

gives way to the upper montane rain forest (zone 7) at approximately 5500 ft (1800 m). This is followed by a distinctly different vegetation zone which represents subalpine open forest and scrub up to about 12 000 ft (3935 m). Here the satellite image offers corrective information, since there is a distinct colour change from dark to light at about 9000 ft (2745 m). This is the upper limit of the mamane (*Sophora chrysophylla*) tree line on the east flank of Mauna Kea.

Note that the northward extension of the montane tropical rain forest on Mauna Kea, as shown on the Ripperton and Hosaka map, does not appear on the satellite image in Figure 2.6. This area has been converted to pasture. It could be that Ripperton and Hosaka ignored this grassland variation and mapped this area as potential montane rain forest, or that the forest conversion took place after 1942. The former seems more probable in light of the next comparison, soil evaluation.

Figure 2.9 shows the soil-order map of the island prepared by Uehara (1973), with minor modifications. Recent lava flows were transferred from the geology map of Abbott (1973) on to the soil map. It becomes evident from comparison of the maps that the soil substrates under Ripperton and Hosaka's vegetation zones are quite variable. In the montane rain forest zone, for example, one finds large areas of histosol (organic soils composed of woody peat on lava bedrock), minor areas of lithosol (prehistoric lava flows), and recent (historic, i.e., <200 years old) lava flows, particularly in the saddle between the two mountains. Further north, much of the east-flank rain forest zone below Mauna Kea is underlain by entisol, i.e., a weakly developed latosol or tropical reddish-brown clay not yet strongly leached of soil nutrients.

The northward extension of Ripperton and Hosaka's montane rain forest zone, which on the satellite image shows an absence of forest, is covered by the same entisol as the forested portion. This supports the contention of Ripperton and Hosaka that the pasture site probably has the potential to support a closed montane rain forest or may revert to rain forest if pasture use is discontinued. On the other hand, the comparison also shows that Ripperton and Hosaka ignored significant substrate variations, such as recent (historic) lava flows. Such substrates are not yet covered by mature rain forest. Instead, they are covered with open woody vegetation, often only scrub. Therefore, Ripperton and Hosaka's vegetation zones are not quite equivalent to Krajina's (1963) biogeoclimatic zones of Hawaii, which are somewhat more narrowly defined landscape units.

With respect to the interpretation of the satellite image, we can now be certain that the dark-coloured patch on the east flank of Mauna Kea is a tropical montane rain forest. This forest is not uniform in plant biomass, nor is it uniform in terms of primary production. Its primary production rate varies with the major substrate types and is clearly lowest on the recent lava flows and next lowest on the lithosols. For a more precise analysis of plant biomass and primary production, one must go from an intermediate to a large-scale map.

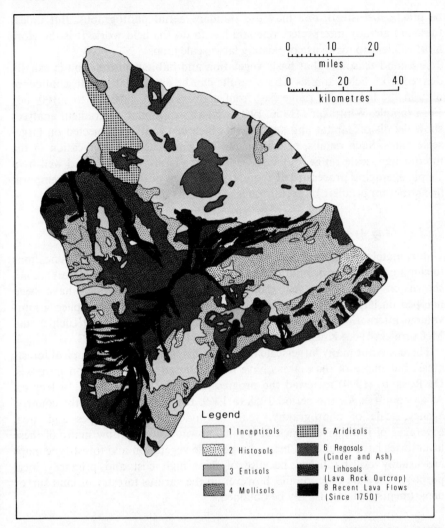

Figure 2.9   Soil order map of the Island of Hawaii at 1:1.3 million after Uehara (1973), modified. Compare with vegetation map, Figure 2.1, p. 33

2.5.2   Accessory Information at the Large-scale Level: the Fine Structure

Large-scale information is an aid to mapping from satellite images for two reasons. One is that the broader landscape types identified on satellite images may often show tonal and textural within-variations, the meaning of which are very difficult to interpret. But it should not always be necessary to do so by acquisition of ground truth. In fact, doing so may not be very helpful. The enlargement from satellite image to field environment may prove too great to be practical. Instead, one may use standard aerial photographs. But unless these are already interpreted, one still has to do the field work. It is therefore most efficient to make use of existing large-scale maps.

A second reason is that basic vegetation and habitat information is usually derived from field samples whose results can be extrapolated with confidence only in the large-scale range (i.e., from approximately 1:10 000 to 1:100 000). For example, Whittaker's (1956, 1960) well-known mountain gradient analyses show details of habitat and vegetation which can only be projected on large-scale maps. Such details, of course, can be generalized for application in the intermediate-scale range, but such generalizing is not accomplished well by a simple averaging procedure. It requires evaluation of the spatial and temporal fine structure peculiar to each broader landscape unit or vegetation zone.

2.5.2.1 Map Availability

Unfortunately, most ecological vegetation and habitat classifications (including vegetation ordinations and environmental gradient analyses) are not shown on maps. However, there are numerous areas which have been mapped on large scales, particularly in Europe. Küchler has compiled a four-volume international bibliography of existing vegetation maps (Küchler and McCormick, 1965; Küchler, 1966, 1968, 1970, 1980).

There are not many large-scale ecological maps available for tropical forests areas, but many of these areas have been assessed for exploitation purposes. De Rosayro (1974) reviewed the progress of forest inventories in 14 tropical Asian countries for the period 1958 to 1969. His tabulation gives the country, agency, scale of photography, year, methodology, forest types and areal coverages of these inventories. It is not known, however, how many of these inventories have resulted in maps. Large-scale vegetation and forest-type maps are usually not published because of their high cost and primarily local purpose. They are often on file, however, in the various forestry or land survey departments and can usually be obtained.

2.5.2.2 Map Interpretation

Generally speaking, large-scale vegetation maps are not easy to interpret. Among other things, they sometimes lack ready information on the location of the mapped area itself. This deficiency can easily be overcome by adding a small insert map that shows the large-scale map area within a continent, subcontinent or island.

A second and more important problem is that most maps are two-dimensional. They do not portray the third dimension easily unless they are specially prepared topographic maps, such as the map in Figure 2.10. Such maps, however, are single-purpose maps, and one cannot show much other information on them.

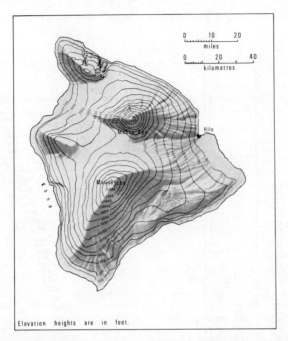

Figure 2.10   Topographic map of the Island of Hawaii at 1:1.2 million. Compare with satellite image shown on Figure 2.7, p. 68

A useful key to map interpretation for ecological purposes is a topographic ecosystem profile, shown in Figure 2.11. This profile diagram was prepared from vegetation map information (obtained through ground truthing) projected initially on 1:12 000 aerial photographs. The completed map was then reduced to 1:48 000. The profile diagram was extracted from this map and from other large-scale topographic, soil and climatic maps. The diagram here is further reduced to a scale of 1:217 000, which is in the intermediate-scale range. The diagram information, however, was based on extrapolation of actual field samples, and the diagram therefore closely reflects the actual field situation. (The profile diagram represents transect 6 on the climate-diagram map in Figure 2.8.)

Profile diagrams of this sort allow one to portray the major ecosystem components—climate, topography, vegetation and soil—in relation to one another. There is not always a clear correspondence of climatic type and vegetation type, or of vegetation with substrate, as can be seen in Figure 2.11.

With the help of Ripperton and Hosaka's (1942) vegetation map of Hawaii it was possible to delineate the montane tropical rain forest on the satellite image. The profile diagram indicates a further breakdown into three types, numbers 8, 9 and 10 on Figure 2.11. Type 9 is an open-structured rain forest which exhibits a high proportion of crown loss due to canopy dieback. The

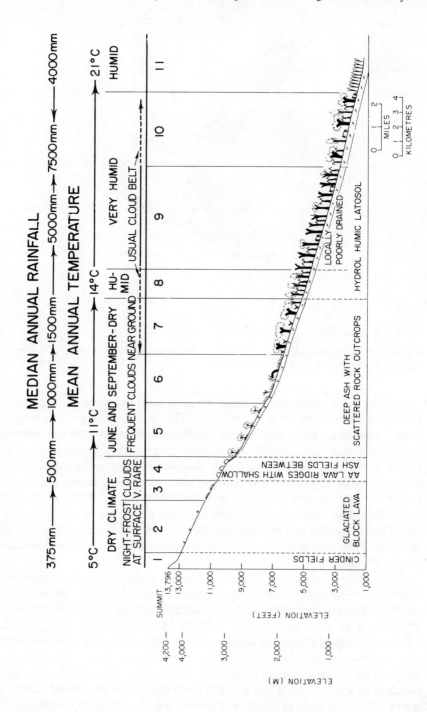

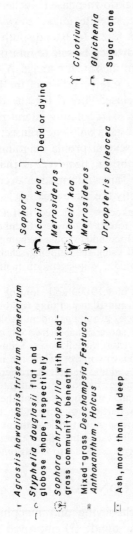

Figure 2.11 Topographic ecosystem profile relating to east-slope of Mauna Kea, Transect 6 on Figure 2.8, p. 70. Ecosystem type names are as follows: 1. Alpine stone desert with crustose lichens, 2 *Agrostis–Trisetum* grass desert, 3 *Styphelia* low-scrub desert, 4 Scattered, globose *Styphelia* scrub, 5 Scattered *Sophora* trees with herbaceous fog-drip communities (treeline vegetation), 6 Grassland with scattered, dying *Sophora* and *Acacia koa* trees, 7 *Acacia koa–Metrosideros–Dryopteris paleacea* forest, 8 Closed *Metrosideros–Cibotium* forest, 9 *Metrosideros–Cibotium–Dicranopteris* forest with dead and dying *Metrosideros* patches, 10 *Acacia koa–Metrosideros–Cibotium–Dicranopteris* forest, 11 Cultivated sugar cane (from Mueller-Dombois and Krajina, 1968)

satellite image shows a lighter shade in the dark-coloured part on the east flank of Mauna Kea which appears to correspond to this dieback forest, which can be interpreted as a natural breakdown phase (Mueller-Dombois *et al.*, 1977; Mueller-Dombois, 1980a). The satellite image shows this lighter shade pattern in the montane tropical rain forest to extend south into Hawaii Volcanoes National Park. It would not, however, be safe to interpret all of this as dieback forest or breakdown phase from the satellite image. A forest can be open-structured for many reasons. A safer interpretation of the lighter shade pattern is simply 'open forest' (versus 'closed forest' for the darker shade pattern). Yet one must also be cautious in extrapolating lateral patterns from profile diagrams. The profile diagram represents only one transect. For a more accurate interpretation, one either needs several profile diagrams or a large-scale map. A third problem of interpretation, however, is that large-scale vegetation maps do not always give the same information even if they are prepared at exactly the same scale for exactly the same area.

At the 1954 World Forestry Congress, Sukachev suggested that his method of mapping should become the standard one for forestry purposes (Ellenberg, 1967). De Phillippis made a counter-proposal that several methods should be compared. Later, a contest to map 420 ha in Switzerland was held using the mapping schemes of Braun-Blanquet, Aichinger, Schmid, and the Eberswalder soil-vegetation method (Mueller-Dombois and Ellenberg, 1974).

The resulting maps, all prepared at a scale of 1:10000 and showing some general resemblances, were all different in detail. The Braun-Blanquet map was based primarily on the mosaic of natural and semi-natural species groupings indentifiable at this large scale. The Aichinger map was based on the successional behaviour of certain key species in relation to types of habitat and silvicultural influences. Schmid's map was based on plant synusiae and the floristic province concept, while the map using the Eberswalder approach was based primarily on soil and habitat characteristics, treating plant species only as indicators of habitat where they had obvious indicator value.

This shows that the user of a vegetation map must be aware of the purposes for which a particular map was prepared. It is particularly important to distinguish between vegetation maps which aim at portraying the intrinsic patterns of vegetation and those which use vegetation to indicate site capability. Both types of maps are useful, but they cannot be used for the same purposes. (See also Daubenmire, 1973.)

2.6 SUMMARY AND CONCLUSIONS

This paper has reviewed methods of classifying and mapping terrestrial vegetation with three main objectives in mind: (1) improved accuracy in estimating world phytomass, (2) improved accuracy in extrapolating primary production from field samples and (3) the monitoring of tropical forest loss.

While a great deal of data on phytomass and primary production has been accumulated by the various IBP projects, very little attention has been given to the question of how to extrapolate these organic production estimates at various spatial levels. In other words, ecosystem functioning has received most of the attention in research funding, while ecosystem structure has been neglected. But questions about the global carbon cycle have underlined the need to focus more attention on ecosystem structure, particularly in tropical forest regions.

Our knowledge of ecosystem structure is now at about the same stage as our knowledge of ecosystem functioning prior to IBP. That is, the methodology for evaluating ecosystem structure from detailed to general and vice versa is quite well worked out except in the tropics at the large-scale level. In contrast, there is a considerable amount of knowledge available on other regions, particularly the northern temperate areas. This information needs to be synthesized, region by region, to deal with the global carbon question. Moreover, research on ecosystem structure in tropical forest areas needs to advance rapidly to match the functional research now under way in the Man-and-The-Biosphere projects concerned with tropical forests. Practically nothing has been done as yet at the large-scale map level (see also Whitmore, 1975; Ashton, 1978; Letouzey, 1978).

2.7 REFERENCES AND BIBLIOGRAPHY

Abbott, A. T. (1973) Geology map of Hawaiian Islands. In: Armstrong, R. W. (ed), *Atlas of Hawaii*, 34–35. University Press of Hawaii, Honolulu. 221 pp.

Ashton, P. S. (1964) Ecological studies in mixed dipterocarp forests of Brunei State. *Oxford Forestry Memoirs, No. 25.* 75 pp.

Ashton, P. S. (1965) Notes on the formation of a rational classification of floristic and structural variation within the mixed dipterocarp forests of Sarawak and Brunei, for forestry and land use planning. *Symposium on Ecological Research in Humid Tropics Vegetation (1963)*, sponsored by Gov. of Sarawak and UNESCO Science Corporation, Office for SE Asia, 185–197. Tokyo Press Co. Ltd, Itabashi, Tokyo. 376 pp.

Ashton, P. S. (1978) The natural forest: plant biology, regeneration and tree growth. *Tropical Forest Ecosystems: A State-of-Knowledge Report prepared by UNESCO/ UNEP/FAO.* UNESCO Natural Resources Research, Vol. XIV, 180–215. 683 pp.

Austin, M. P., Ashton, P. S., and Greig-Smith, P. (1972) The application of quantitative methods to vegetation survey. III. A reexamination of rain forest data from Brunei. *Journal of Ecology*, **60**, 305–332.

Barbour, M. G., Burk, J. H., and Pitts, W. D. (1980) *Terrestrial Plant Ecology*. The Benjamin Cummings Publishing Co., Menlo Park, California, London, Sidney. 604 pp.

Barkman, J. J. (1978) Synusial approaches to classification. In: Whittaker, R. H. (ed), *Classification of Plant Communities*, 111–165. Junk Publishers, The Hague. 408 pp.

Barney, G. (1978) The nature of the deforestation problem—trends and policy implications. *Proceedings of the US Strategy Conference on Tropical Deforestation*, 15–22. US State Dept. and US Agency for International Development Publication, Washington, DC. 78 pp.

Baumgartner, A., and Brünig, E. F. (1978) Tropical forests and the biosphere. *Tropical Forest Ecosystems: A State-of-Knowledge Report Prepared by UNESCO/ UNEP/FAO.* UNESCO Natural Resources Research, Vol. XIV, 33–60. 683 pp.

Beals, E. (1960) Forest bird communities in the Apostle Islands of Wisconsin. *Wilson Bulletin*, **72**, 156–181.

Beard, J. S. (1946) The Mora forests in Trinidad, British West Indies. *Journal of Ecology*, **33**, 173–192.

Beard, J. S. (1955) The classification of tropical American vegetation types. *Ecology*, **36**(1), 89–100.

Beard, J. S. (1978) The physiognomic approach. In: Whittaker, R. H. (ed), *Classification of Plant Communities*, 33–64. Junk Publishers, The Hague. 408 pp.

Becker, R. B. (1976) The phytosociological position of tree ferns (*Cibotium* spp.) in the montane rain forests on the Island of Hawaii. PhD dissertation, University of Hawaii. 368 pp.

Blasco, F. (1971) Montagnes du sud de l'Inde: forêsts, saveanes, écologie. Institut Francais de Pondichery. *Travaux de la Section Scientifique et Technique*. Tome X, Fasc. 1. 436 pp.

Bolin, B. (1977) Changes of land biota and their importance for the carbon cycle. *Science*, **196**(4290), 613–615.

Box, E. (1975) Quantitative evaluation of global primary productivity models generated by computers. In: Lieth, H. and Whittaker, R. H. (eds), *Primary Productivity of the Biosphere. Ecology Studies*, **14**, 265–283. Springer Verlag, New York. 339 pp.

Braun-Blanquet, J. (1928) *Pflanzensoziologie*. 2nd ed. 1951, 3rd ed. 1964. Springer Verlag, Wien, New York. 865 pp.

Braun-Blanquet, J. (1965) *Plant Sociology: The Study of Plant Communities*. (Translated, revised and edited by Fuller, C. D. and Conard, H. S.) Hafner Publishing Co., London. 439 pp.

Bray, J. R., and Curtis, J. T. (1957) An ordination of the upland forest communities of southern Wisconsin. *Ecological Monographs*, **27**, 325–349.

Brockmann-Jerosch, H., and Rübel, E. (1912) Die Einteilung der Pflanzengesellschaften nach ökologisch-physiognomischen. Gesichtspunkten. Verlag Wilhelm Engelmann, Leipzig. 68 pp.

Brünig, E. F. (1970) Stand structure, physiognomy and environmental factors in some lowland forests in Sarawak. *Tropical Ecology*, **11**(1), 26–43.

Burtt-Davy, J. (1938) The classification of tropical woody vegetation types. *Oxford Imp. For. Inst. Paper No. 13.* 85 pp.

Cajander, A. K. (1909) Über Waldtypen. *Acta Forestalia Fennica*, **1**, 1–125.

Cêska, A., and Roemer, H. (1971) A computer program for identifying species-relevé groups in vegetation studies. *Vegetatio*, **23**(3–4), 255–277.

Champion, H. G. (1936) A preliminary survey of the forest types of India and Burma. *Indian Forest Records N. S. Silviculture*, **1**, 1–287.

Clements, F. E. (1916) *Plant Succession. An Analysis of the Development of Vegetation.* Carnegie Institute, Washington, DC. 512 pp.

Clements, F. E. (1928) *Plant Succession and Indicators.* H. W. Wilson Co., New York, London. 453 pp.

Cooray, R. G. (1974) Stand structure of a montane rain forest on Mauna Loa, Hawaii. Island Ecosystems IRP/IBP Hawaii (University of Hawaii). *Technical Report 44.* 98 pp.

Cottam, G., Goff, F. G., and Whittaker, R. H. (1978) Wisconsin comparative ordination. In: Whittaker, R. H. (ed), *Ordination of Plant Communities*, 185–213. Junk Publishers, The Hague. 388 pp.

Dagnelie, P. (1960) Contribution a l'étude des communautés végétales par l'analyse factorielle. *Bulletin Service Carte Phytogeogr.*, Sér. B (5), 7–71 and 93–175.

Dagnelie, P. (1978) Factor analysis. In: Whittaker, R. H. (ed), *Ordination of Plant Communities*, 215–238. Junk Publishers, The Hague. 388 pp.

Dansereau, P. (1951) Description and recording of vegetation upon a structural basis. *Ecology*, **32**, 172–229.

Dansereau, P. (1957) *Biogeography, An Ecological Perspective.* Ronald Press, New York. 394 pp.

Dansereau, P. *et al.* (1968) The continuum concept of vegetation: responses. *Botanical Review*, **34**(3), 253–332.

Dasmann, R. F. (1973) A system for defining and classifying natural regions for purposes of conservation. *IUCN Occas. Paper No. 7.* 48 pp.

Daubenmire, R. F. (1968) *Plant Communities: A Textbook of Plant Synecology.* Harper and Row, New York, Evanston, London. 300 pp.

Daubenmire. R. F. (1973) A comparison of approaches to the mapping of forest land for intensive management. *Forestry Chronicle*, **49**(2), 87–91.

Davis, T. A. W., and Richards, P. W. (1933–4) The begetation of Morabilli Creek, British Guiana: An ecological study of a limited area of tropical rain forest. *Journal of Ecology*, **21**, 350–384; **22**, 106–155.

Diels, L., and Mattick, F. (1908) *Pflanzengeographie*. Samml. Göschen Nr. 389–389a. 1958: Walter du Gruyter, Berlin, Leipzig. 160 pp.

Driscoll, R. S., and Spencer, M. M. (1972) Multispectral scanner imagery for plant community classification. *Proceedings of the 8th International Symposium on Remote Sensing of Environment*, 1259–1278. Ann Arbor, Michigan.

Drude, O. (1902) Grundzüge der Pflanzenverbreitung. In: Engler, A. and Drude, O. (eds), *Die Vegetation der Erde*, Vol. VI. Verlag Wilh. Engelmann, Leipzig. 671 pp.

Du Rietz, G. E. (1921) *Zur methodischen Grundlage der modernen Pflanzensoziologie.* Akadem. Abh. U0sala. 272 pp.

Ellenberg, H. (1950) *Unkraut-Gemeinschaften als Zeiger für Klima und Boden.* Verlag Eugen Ulmer, Stuttgart. 141 pp.

Ellenberg, H. (1952) *Landwirtschaftliche Pflanzensoziologie*, Bd. II: Wiesen und Weiden und ihre standörtliche Bewertung, Ludwigsburg. 143 pp.

Ellenberg, H. (ed) (1967) *Vegetations- und bodenkundliche Methoden der forstlichen Standortskartierung.* Veröffentl. geobot. Inst. Rübel, Zürich, Heft 39. 298 pp.

Ellenberg, H. (1973) Die Ökosysteme der Erde: Versuch einer Klassifikation der Ökosysteme nach funktionalen Gesichtspunkten. In: Ellenberg, H. (ed), *Ökosystemforschung*, 235–265. Springer-Verlag, Berlin, Heidelberg, New York. 280 pp.

Ellenberg, H. (1978) *Vegetation Mitteleuropas mit den Alpen.* 2nd ed. Ulmer Verlag, Stuttgart. 981 pp.

Ellenberg, H., and Mueller-Dombois, D. (1967a) Tentative physiognomic–ecological classification of plant formations of the earth. *Ber. geobot. Inst. ETH, Stiftg. Rübel, Zürich*, **37**, 31–55.

Ellenberg, H., and Mueller-Dombois, D. (1967b) A key to Raunkiaer plant life forms with revised subdivisions. *Ber. geobot. Inst. ETH Stiftg. Rübel, Zürich*, **37**(1965/66), 56–73.

Es, E. van (1976) The dry deciduous forest of Bastar, central India, on LANDSAT-1. *ITC Journal*, **2**, 332–340.

Estes, J. E., and Simonett, D. S. (eds) (1975) Fundamentals of image interpretation. In: Reeves, R. G., Anson, A., and Landen, D. (eds), *Manual of Remote Sensing*, Vol. II, 869–2144. American Society of Photogrammetry, Falls Church, Virginia.

Ewel, J. J., Madriz, A., and Tosi, J. A. (Jr.) (1968) *Zonas de Vida de Venezuela.* Republica de Venezuela, Ministerio de Agricultura y Cria Direcion de Investigacion. 264 pp + map.

Ewel, J. J., and Whitmore, J. L. (1973) The ecological life zones of Puerto Rico and the Virgin Islands. USDA Forest Service, Institute of Tropical Forestry, Puerto Rico, *Forest Service Research Paper ITF 18.* 72 pp.

FAO (1975) Formulation of a tropical forest cover monitoring project. *FAO Publication.* Rome. 76 pp.

Fontaine, R. G. (1978) Inventory and survey: international activities. *Tropical Forest*

Ecosystems: A State-of-Knowledge Report Prepared by UNESCO/UNEP/FAO. UNESCO Natural Resources Research, Vol. XIV, 17–32. 683 pp.

Fosberg, R. F. (1961) A classification of vegetation for general purposes. *Tropical Ecology*, **2**(1,2), 1–28.

Fosberg, R. F. (1967) A classification of vegetation for general purposes. Guide to the check sheet for IBP areas by G. F. Peterken. *IBP Handbook No. 4*, 73–102. Blackwell Scientific Publications, Oxford, Edinburgh. 133 pp.

Gams. H. (1918) Prinzipienfragen der Vegetationsforschung. Ein Beitrag zur Begriffsklärung und Methodik der Biocoenologie. *Vierteljahrsschr. Naturforsch. Ges. Zürich*, **63**, 293–493.

Gaussen, H. (1954) Théories et classification des climats et des microclimats du point de vue photogeographique. In *8e Cong. Intern. Botan. (Paris), Section 7*, 125–130.

Gaussen, H. (1957) Integration of data by means of vegetation maps. *Proceed. 9th Pacific Sci. Congress, Bangkok.*

Gaussen, H. (1964) Les cartes de végétation. Institut Francais de Pondichéry. *Travaux Section Scientifique Technique T. I. Fasc. 2.*

Gaussen, H., Legris, P., and Viart, M. (1961) Notice de la feuille Cape Comorin. Carte Internationale du Tapis Végétal a 1/1,000,000 Inst. fr. Pondichery. *Travaux Section Scientifique Technique Hors. Série No. 1.* 108 pp.

Gaussen, H., Legris, P., Viart, M., and Labroue, L. (1964) *International Map of the Vegetation, Ceylon.* Special sheet published by the Ceylon (now Sri Lanka) Survey Department.

Gaussen, H., Legris, P., Viart, M., and Labroue, L. (1965) *Notice de la feuille, Ceylon.* De l'Institut Francais de Pondichéry, India, Hors Série No. 5, 78 pp.

Gaussen, H., Legris, P., and Blasco, F. (1967) *Bioclimats du Sud-Est Asiatique.* Institut Francais de Pondichéry. Tome III, Fasc. 4, 114 pp.

Gerrish, G., and Mueller-Dombois, D. (1980) Behavior of native and non-native plants in two tropical rain forests on Oahu, Hawaiian Islands. *Phytocoenologia*, **8**, 237–295.

Gleason, H. A. (1926) The individualistic concept of the plant association. *Bulletin Torrey Botanical Club*, **53**, 7–26.

Goff, F. G., and West, D. (1975) Canopy-understory interaction effects on forest population structure. *Forest Science*, **21**, 98–108.

Graebner, P. (1925) Die Heide Norddeutschlands. In: Engler, A. and Drude, O. (eds), *Die Vegetation der Erde*, Vol. V, 2nd ed. Verlag Wilh. Engelmann, Leipzig. 277 pp.

Grisebach, A. (1872) *Die Vegetation der Erde nach ihrer klimatischen Anordnung.* Verlag Wilh. Engelmann, Leipzig, Vol. 1, 602 pp; Vol. 2, 635 pp.

Hett, J. M., and Loucks, O. L. (1971) Sugar maple (*Acer saccharum* Marsh) seedling mortality. *Journal of Ecology*, **59**, 507–520.

Hill, M. O. (1973) Reciprocal averaging: An eigenvector method of ordination. *Journal of Ecology*, **61**, 237–249.

Holdridge, L. R. (1947) Determination of world formations from simple climatic data. *Science*, **105**, 367–368.

Holdridge, L. R. (1967) *Life Zone Ecology.* Tropical Science Center, San Jose, Costa Rica. 206 pp.

Hueck, K. (1966) *Die Wälder Südamerikes.* Fischer Verlag, Stuttgart. 422 pp.

Hueck, K., and Seibert, P. (1972) *Vegetationskarte von Südamerika.* Gustav Fischer Verlag, Stuttgart.

Jobin, K., and Beaubien, J. (1974) Capability of ERTS-1 imagery for mapping forest cover types of Anticosti Island. *Forestry Chronicle*, **50**(6), 233–237.

Kartawinata, K. (1977) The ecological zones of Indonesia. In: *Papers presented at the*

13th Pacific Science Congress, Vancouver, Canada, 1975, 51–58. Lembago Ilmu Pengetahuan, Indonesia.

Knapp, R. (1973) *Die Vegetation von Afrika*. G. Fischer Verlag, Stuttgart. 626 pp.

Knight, D. H. (1975) A phytosociological analysis of species-rich tropical forest on Barro Colorado Island, Panama. *Ecological Monographs*, **45**, 259–284.

Knight, D. H., and Loucks, O. L. (1969) A quantitative analysis of Wisconsin forest vegetation on the basis of plant function and gross morphology. *Ecology*, **50**, 219–234.

Köppen, W. (1931) Grundriss der Klimakunde. De Gruyter Verlag, Berlin.

Köppen, W. (1936) Das geographischa System der Klimate. In: Köppen, W. and Geiger, G. (eds), *Handbuch der Klimatologie*, Vol. 1, Part C. Gebr. Borntraeger, Berlin.

Krajina, V. J. (1960) Can we find a common platform for the different schools of forest-type classification? *Silva Fennica, Helsinki*, **105**, 50–59.

Krajina, V. J. (1963) Biogeoclimatic zones of the Hawaiian Islands. *Hawaiian Botanical Society Newsletter*, **2**(7), 93–98.

Krajina, V. J. (1965) Biogeoclimatic zones and biogeocoenoses of British Columbia. *Ecology of Western North America. Department of Botany, University of British Columbia*, **1**, 1–17.

Krajina, V. J. (1969) Ecology of forest trees in British Columbia. *Ecology of Western North America. Department of Botany, University of British Columbia*, **2**, 1–147.

Krajina, V. J. (1973) Biogeoclimatic zonation as a basis for regional ecosystems. *Symposium on Planned Utilization of Lowland Tropical Forests*. Cipayung, Bogor 1971, 18–32.

Krajina, V. J. (1974) *Biogeoclimatic zones of British Columbia. Map at 1:1.9 million*. British Columbia Ecological Reserves Committee, Department of Lands, Forests and Water Resources, Victoria, BC.

Küchler, A. W. (1947) A geographic system of vegetation. *Geographical Review*, **37**, 233–240.

Küchler, A. W. (1949) A physiognomic classification of vegetation. *Annals, Association American Geographers*, **39**, 201–210.

Küchler, A. W. (1964) Potential natural vegetation of the conterminous United States. Manual to accompany the map. *American Geographic Society, Special Publication No. 36*. 116 pp.

Küchler, A. W. (1965) Potential natural vegetation map at 1:7.5 million. (Revised ed.) *USDA Geological Survey, Sheet No. 90*.

Küchler, A. W. (1966) International bibliography of vegetation maps. *Vol. II: Vegetation Maps of Europe*. University of Kansas Library Series. 584 pp.

Küchler, A. W. (1967) *Vegetation Mapping*. The Ronald Press Company, New York. 472 pp.

Küchler, A. W. (1968) International bibliography of vegetation maps. *Vol. III: U.S.S.R., Asia and Australia*. University of Kansas Library Series. 389 pp.

Küchler, A. W. (1970) International bibliography of vegetation maps. *Vol. IV. Vegetation Maps of Africa, South America and the World (General)*. University of Kansas Library Series. 561 pp.

Küchler, A. W. (1972) On the structure of vegetation. *Ecology*, **53**, 196–198.

Küchler, A. W. (1974) A new vegetation map of Kansas. *Ecology*, **55**, 586–604.

Küchler, A. W. (1980) International bibliography of vegetation maps, 2nd ed., Sect 1: *South America*. University of Kansas Library Series No. 45.

Küchler, A. W., and McCormick, J. (1965) International bibliography of vegetation maps. *Vol. 1: North America*. University of Kansas Library Series. 453 pp.

Küchler, A. W., and Montoya-Maquin, J. M. (1971) The UNESCO classification of vegetation: Some tests in the tropics. Instituto Inter-Americano de liencias Agricolas. *Turrialba*, **21**, 98–109.

Küchler, A. W., and Sawyer, J. O., Jr. (1967) A study of the vegetation near Chiengmai, Thailand. *Transactions, Kansas Academy of Science*, **70**(3), 281–348.

Lawrence, R. D., and Herzog, J. H. (1975) Geology and forestry classification from ERTS-1 digital data. *Photogrammetric Engineering and Remote Sensing*, **41**(10), 1241–1251.

Leak, W. B. (1965) The J-shaped probability distribution. *Forest Science*, **11**, 405–409.

Letouzey, R. (1978) Floristic composition and typology. *Tropical Forest Ecosystems: A State-of-Knowledge Report Prepared by UNESCO/UNEP/FAO*. 91–111, UNESCO Natural Resources Research, Vol. XIV. 693 pp.

Lieth, H. (1964) Versuch einer kartographischen Erfassung der Stoffproduktion der Erde. *Geographisches Taschenbuch 1964/65*, 72–80. Steiner Verlag, Wiesbaden.

Lieth, H. (1973) Primary production: Terrestrial ecosystems. *Human Ecology*, **1**(4), 303–332.

Lieth, H. (1975) Primary productivity in ecosystems: Comparative analysis of global patterns. In: Dobben, W. H. van and Lowe-McConnel, R. H. (eds), *Unifying Concepts in Ecology*, 67–88. Junk Publishers, The Hague. 302 pp.

Lieth, H., and Moore, G. W. (1971) Computerized clustering of species in phytosociological tables and its utilization for field work. Spatial Patterns and Statistical Distributions. *Statistical Ecology, Vol. I*, 403–422. Pennsylvania State University Press.

Lieth, H., and Whittaker, R. H. (eds) (1975) Primary Productivity of the Biosphere. *Ecological Studies, 114*. Springer-Verlag, New York, Heidelberg, Berlin. 339 pp.

Lippman, T. (1939) The unistratal concept of plant communities (the unions). *American Midland Naturalist*, **21**, 111–145.

Loucks, O. L. (1962) Ordinating forest communities by means of environmental scalars and phytosociological indices. *Ecological Monographs*, **32**, 137–166.

McIntosh, R. P. (1967) The continuum concept of vegetation. *Botanical Review*, **33**(2), 130–187.

McIntosh, R. P. (1978) Matrix and plexus techniques. In: Whittaker, R. H. (ed), *Ordination of Plant Communities*, 152–184. Junk Publishers, The Hague, 387 pp.

Meyer, H. A. (1952) Structure, growth and drain in uneven-aged forests. *Journal of Forestry*, **50**, 85–92.

Mohr, E. C. J. (1933) De bodem der tropen in het algemeen en die van Nederlandsch-Indie in het bijzonder. *Kon. Ver. Kolon Instit., Amsterdam Handelsmuseum, Meded.*, **21**, 91–110.

Motyka, J., Dyobrzanski, B., and Zawadski, S. (1950) Wstepne badania nad lakami poludniowo-wschodneij Lubelszczyzny. *Annls. University M. Curie-Sklodowska, Sec. E.*, **5**(13), 367–447.

Mueller-Dombois, D. (1964) The forest habitat types in southeastern Manitoba and their application to forest management. *Canadian Journal of Botany*, **42**, 1417–1444.

Mueller-Dombois, D. (1965) Eco-geographic criteria for mapping forest habitats in southeastern Manitoba. *Canadian Forestry Chronicle*, **41**(2), 188–206.

Mueller-Dombois. D. (1966a) Climate. In: Doty, M. S. and Mueller-Dombois, E. (eds), *Atlas for Bioecology Studies in Hawaii Volcanoes National Park*, 85–92. College of Tropical Agriculture, Hawaii Agriculture Experiment Station, Bulletin 89. 507 pp.

Mueller-Dombois, D. (1966b) The vegetation map and vegetation profiles. In: Doty, M. S. and Mueller-Dombois, D. (eds), *Atlas for Bioecology Studies in Hawaii Volcanoes National Park*, 391–441. College of Tropical Agriculture, Hawaii Agriculture Experiment Station, Bulletin 89. 507 pp.

Mueller-Dombois, D. (1968) Eco-geographic analysis of a climate map of Ceylon with particular reference to vegetation. *Ceylon Forester*, **8**(3/4), 39–54 + map.

Mueller-Dombois, D. (1969) *Vegetation map of Rubuna National Park*. Five sheets at 1:31,680. Survey Department of Sri Lanka.

Mueller-Dombois, D. (1970) *Vegetation map of Wilpattu National Park*. Four sheets at 1:31,680. Survey Department of Sri Lanka.

Mueller-Dombois, D. (1980a) The 'Ōhi'a dieback phenomenon in the Hawaiian rain forest. In: Cairns, J. Jr. (ed), *The Recovery Process in Damaged Ecosystems*, 153–161. Ann Arbor Science Publishers, Inc., Ann Arbor. 167 pp.

Mueller-Dombois, D. (1980b) Fire in tropical ecosystems. pp. 137–176. In: Mooney, H., Bonnicksen, T. M., and Christensen, N. L. (eds), *Fire Regimes and Ecosystem Properties*. USDA Forest Service, General Technical Report WO-26.

Mueller-Dombois, D., Bridges, K. W., and Carson, H. L. (eds) (1981) *Island Ecosystems: Biological Organization in Selected Hawaiian Communities*. US/IBP Synthesis Series 15. Hutchinson Ross Publishing Company, Stroudsburg, Pennsylvania. 583 pp.

Mueller-Dombois, D., and Ellenberg, H. (1974) *Aims and Methods of Vegetation Ecology*. John Wiley and Sons, New York. 547 pp.

Mueller-Dombois, D., and Ellenberg, H. (1980) Methods available for classification and their suitability for various purposes. In: Clapham, A. R. (ed). *The IBP Survey of Conservation Sites: An Experimental Study*, 26–54. International Biological Programme 24. Cambridge University Press, Cambridge, London, New York. 344 pp.

Mueller-Dombois, D., and Fosberg, F. R. (1974) Vegetation map of Hawaii Volcanoes National Park (at 1:52,000). *CPSU Technical Report No. 4*. Botany Department, University of Hawaii. 44 pp.

Mueller-Dombois, D., Jacobi, J. D., Cooray, R. G., and Balakrishnan, N. (1977) 'Ōhi'a a rain forest study: Ecological investigations of the 'ōhi'a dieback problem in Hawaii. *Coop. National Parks Studies Unit Technical Report No. 20*. Botany Department, University of Hawaii. 117 pp + 3 maps.

Mueller-Dombois, D., and Krajina, V. J. (1968) Comparison of eastflank vegetations on Mauna Loa and Mauna Kea, Hawaii. *Recent Advances in Tropical Ecology*, II, 508–520.

Orloci, L. (1966) Geometric models in ecology. I. The theory and application of some ordination methods. *Journal of Ecology*, **54**, 193–215.

Orloci, L. (1967) An agglomerative method to classify plant communities. *Journal of Ecology*, **55**, 193–205.

Orloci, L. (1978) *Multivariate Analysis in Vegetation Research*. Junk Publishers, The Hague. 451 pp.

Poore, M. E. D. (1955) The use of phytosociological methods in ecological investigations. I. The Braun-Blanquet System. *Journal of Ecology*, **43**, 226–244. II. Practical issues involved in an attempt to apply the Braun-Blanquet system. *Journal of Ecology*, **43**, 245–269. III. Practical applications. *Journal of Ecology*, **43**, 606–651.

Poore, M. E. D. (1964) Integration in the plant community. *Journal of Ecology*, **52** (Supplement), 213–226.

Poore, M. E. D. (1968) Studies in Malaysian rain forest. I. The forest of Triassic sediments in Jengka Forest Reserve. *Journal of Ecology*, **56**, 143–196.

Poore, M. E. D. (1978) Tropical rain forests and moist deciduous forests. *IUCN General Assembly Paper GA 78/10 Add. 1*. 30 pp.

Raunkiaer, C. (1937) *Plant Life Forms*. Clarendon, Oxford. 104 pp.

Reeves, R. G., Abraham, A., and Landen, D. (eds) (1975) *Manual of Remote Sensing*, Vol. II, 869–2144. American Society Photogrammetry, Falls Church, Virginia.

Richards, P. W. (1952) *The Tropical Rain Forest*. University Press, Cambridge. 450 pp.

Richards, P. W., Tansley, A. G., and Watt, A. S. (1940) The recording of structure, life form and flora of tropical forest communities as a basis for their classsification. *Journal of Ecology*, **28**, 224–239.

Ripperton, J. C., and Hosaka, E. Y. (1942) Vegetation zones of Hawaii. *Hawaii Agricultural Experiment Station Bulletin No. 89*. 60 pp + 3 maps.

Rodin, L. E., Bazilevich, N. I., and Rozov, N. N. (1975) Productivity of the world's main ecosystems. Productivity of World Ecosystems. *Proceedings, Symposium Fifth General Assembly of Special Committee of IBP*, 13–26. National Academy of Sciences, Washington, DC. 166 pp.

Rosayro, R. A. de (1974) Vegetation of humid tropical Asia. *Natural Resources Research*, XII, 179–195. UNESCO Publication, Paris.

Rübel, E. (1933) Geographie der Pflanzen (Soziologie). *Handwörterbuch d. Naturwiss. Jena*, 1044–1971.

Schimper, A. F. W. (1898) *Pflanzengeographie auf ökologischer Grundlage*, 1935 3rd ed. by F. C. v. Faber. Verlag Gustav Fischer, Jena, I, 1–588, II, 589–1612.

Schmid, E. (1963) Die Erfassung der Vegetationseinheiten mit floristischen und epimorphologischen Analysen. Berichte Schweiz. *Botanische Gesellschaft*, **73**, 276–324.

Schmithüsen, J. (1968) 1:25 million vegetation maps of Europe, North Asia, South Asia, SW Asia, Australia, N. Africa, S. Africa, N. America, Central America, South America (north part), South America (south part). In: *Grosses Duden-Lexikon*, Vol. 8, 321–346. Bibliographisches Institut A.G., Mannheim.

Sharpe, D. M. (1975) Methods of assessing primary production of regions. In: Lieth, H. and Whittaker, R. H. (eds), *Primary Productivity of the Biosphere. Ecological Studies*, **14**, 147–166. Springer-Verlag, New York. 339 pp.

Shimwell, D. W. (1971) *The Description and Classification of Vegetation*. University of Washington Press, Seattle. 322 pp.

Sicco Smit, G. (1974) Practical applications of radar images to tropical rain forest mapping in Colombia. *Mitt. Bundesforschungsanstalt für Forst.-u. Holzwirtschaft*, **99**, 51–64 + 1 map.

Sokal, R. R., and Sneath, P. H. A. (1963) *Principles of Numerical Taxomony*. Freeman, San Francisco, London. 359 pp.

Sommer, A. (1976) Attempt at an assessment of the world's tropical forests. *Unasylva*, **28**(112/113), 5–25.

Sørensen, T. (1948) A method of establishing groups of equal amplitude in plant sociology based on similarity of species content. *Det. Kong. Danske Vidensk. Selsk. Biol. Skr. (Copenhagen)*, **5**(4), 1–34.

Spatz, G., and Siegmund, J. (1973) Eine Methode zur tabellarischen Ordination, Klassifikation und ökologischen Auswertung von pflanzensoziologischen Bestandesaufnahmen. *Vegetatio*, **28**, 1–17.

Specht, R. L. (1970) Vegetation. In: Leeper, G. W. (ed), *The Australian Environment*, 4th ed., 44–67. CSIRO, Melbourne University Press, Australia.

Specht, R. L., Roe, E. M., and Boughton, V. H. (eds) (1974) Conservation of Major Plant Communities in Australia and Papua New Guinea. *Austral. Journal Botany, Supplementary Ser. Supplement No. 7*, CSIRO, Melbourne, Australia. 666 pp.

Sukachev. V. N. (1928) Principles of classification of the spruce communities of European Russia. *Journal of Ecology*, **16**, 1–18.

Sukachev, V. N. (1932) Die Untersuchung der Waldtypen des osteuropäischen Flachlandes. *Handbuch der Biologoischen Arbeitsmethoden*, **11**(6), 191–250.

Sukachev, V. N. (1945) Biogeocoenology and phytocoenology. *C.R. Academy of Science U.S.S.R.*, **47**, 429–431.

Sukachev, V. N., and Dylis, N. (1964) *Fundamentals of Forest Biogeocoenology.* Translation by J. M. Maclennan (1968). Oliver and Boyd, Edinburgh, London, 672 pp.

Tansley, A. G., and Chipp, T. F. (eds) (1926) *Aims and Methods in the Study of Vegetation.* The British Empire Vegetation Committee, Whitefriars Press, London. 383 pp.

Thompson, W. Y. (1978) *Report to the Governor, 1976–77.* Department of Land and Natural Resources, State of Hawaii, Honolulu. 73 pp.

Thorley, G. A. (ed) (1975) Forest Lands: Inventory and Assessment, 1353–1426. *Manual of Remote Sensing.* Vol. II, 869–2144. American Society of Photogrammetry, Falls Church, Virginia.

Thornthwaite, C. W. (1948) An approach toward a rational classification of climate. *Geographical Review*, **38**, 55–94.

Thornthwaite, C. W., and Mather, J. R. (1957) Instructions and tables for computing potential evapotranspiration and the water balance. *Publications in Climatology*, **10**(3), 1–311.

Tiwari, K. P. (1976) An approach to regional land use classification using LANDSAT imagery. *Indian Forester*, **102**(11), 791–807.

Tomlinson, P. B., and Gill, A. M. (1973) Growth habits of tropical trees: Some guiding principles. In: Meggers, B. J., Ayensu, E. S., and Duckworth, W. D. (eds), *Tropical Forest Ecosystems in Africa and South America: A Comparative Review*, 129–143. Smithsonian Institute Press, Washington, DC. 350 pp.

Tüxen, R. (1970) Entwicklung, Stand und Ziele der pflanzensoziologischen Systematik (Syntaxonomie). *Berichte der deutschen botanischen Gesellschaft*, **83**, 633–699.

Uehara, G. (1973) Soil order map of Hawaiian Islands. In: Armstrong, R. W. (ed), *Atlas of Hawaii*, 40–41. University Press of Hawaii, Honolulu. 221 pp.

UNESCO (1973) International classification and mapping of vegetation. *UNESCO Ecology and Conservation Series* 6. 93 pp.

Vareschi, V. (1968) Comparación entre selvas neotropiaceles y paleotropicales en base a su espectro de biotipos. *Acta Bontanica Venezuelica*, **3**, 239–263.

Wacharakitti, S. (1975) Automatic analysis of remote sensing data. *Research Note No. 18.* Kasetsart University, Faculty of Forestry, Bangkok, Thailand. 5 pp.

Walter, H. (1955) Die Klimadiagramme als Mittel zur Beurteilung der Klimaverhält-nisse für ökologische, vegetationskundliche und landwirtschaftliche Zwecke. *Berichte der deutschen botanischen Gesellschaft*, **68**, 331–344.

Walter, H. (1957) Wie kann man den Klimatypus anschaulich darstellen? *Die Umschau in Wissenschaft und Technik.* Heft 24, 751–752.

Walter, H. (1971) *Ecology of Tropical and Subtropical Vegetation.* Translation by D. Mueller-Dombois. Oliver and Boyd, Edinburgh. 539 pp.

Walter, H. (1973a) *Vegetation of the Earth.* Translation by Joy Wieser. Springer-Verlag, New York. 237 pp.

Walter, H. (1973b) Die Vegetation der Erde in öko-physiologischer Betrachtung. Band I: *Die tropischen und subtropischen Zonen*, 3rd ed. VEB Fischer Verlag, Jena. 743 pp.

Walter, H. (1976) Die ökologischen Systeme der Kontinente (Biogeosphäre). Prinzipien ihrer Gliederung mit Beispielen. Fischer Verlag, Stuttgart, New York. 131 pp.

Walter, H., Harnickell, E., and Mueller-Dombois, D. (1975) *Climate-diagram maps of the individual continents and the ecological climatic regions of the earth.* Springer-Verlag, New York. 36 pp + 9 maps.

Walter, H., and Lieth, H. (1960) *Klimadiagramm-Weltatlas.* VEB Fischer Verlag, Jena.

Warming, E. (1909) Oecology of Plants. *An Introduction to the Study of Plant Communities.* Oxford University Press. 422 pp.

Webb, L. J. (1959) Physiognomic classification of Australian rain forests. *Journal of Ecology*, **47**, 551–570.

Webb, L. J. (1968) Environmental relationships of the structural types of Australian rain forest vegetation. *Ecology*, **49**(2), 298–311.

Webb, L. J. (1978) A general classification of Australian rain forests. *Australian Plants*, **9**(76), 349–363.

Webb, L. J., Tracy, J. G., Williams, W. T., and Lance, N. G. (1970) Studies in the numerical analysis of complex rain-forest communities. *Journal of Ecology*, **58**, 203–232.

Webb, L. J., Tracy, J. G., and Williams, W. T. (1976) The value of structural features in tropical forest typology. *Australian Journal of Ecology*, **1**, 3–28.

Westhoff, V., and Maarel, E. van der (1978) The Braun-Blanquet Approach. In: Whittaker, R. H. (ed), *Classification of Plant Communities*, 1287–1399. Junk Publishers, The Hague. 408 pp.

Whitmore, T. C. (1975) *Tropical Rain Forests of the Far East*. Clarendon Press, Oxford. 282 pp.

Whittaker, R. H. (1956) Vegetation of the Great Smoky Mountains. *Ecological Monographs*, **26**, 1–80.

Whittaker, R. H. (1960) Vegetation of the Siskiyou Mountains, Oregon and California. *Ecological Monographs*, **30**, 279–338.

Whittaker, R. H. (1962) Classification of natural communities. *The Botanical Review*, **28**(1), 1–239.

Whittaker, R. H. (1967) Gradient analysis of vegetation. *Biologica. Review*, **42**, 207–264.

Whittaker, R. H. (1970) *Communities and Ecosystems. Current Concepts in Biology Series*. The Macmillan Company, London. 162 pp; 2nd ed. (1975). 287 pp.

Whittaker, R. H. (ed) (1978a) *Ordination of Plant Communities*. Junk Publishers, The Hague, Boston. 388 pp.

Whittaker, R. H. (ed) (1978b) *Classification of Plant Communities*. Junk Publishers, The Hague, Boston. 408 pp.

Whittaker, R. H., and Gauch, H. G., Jr. (1978) Evaluation of ordination techniques. In: Whittaker R. H. (ed), *Ordination of Plant Communities*, 277–336. Junk Publishers, The Hague. 388 pp.

Whittaker, R. H., and Likens, G. E. (1975) The biosphere and man. In: Leith, H., and Whittaker, R. H. (eds), *Primary Productivity of the Biosphere. Ecological Studies*, **14**, 305—328. Springer-Verlag, New York, Heidelberg, Berlin. 339 pp.

Whittaker, R. H., and Niering, W. A. (1965) Vegetation of the Santa Catalina Mountains, Arizona: A gradient analysis of the south slope. *Ecology*, **46**(4), 429–452.

Wong, C. S. (1978) Atmospheric input of carbon dioxide from burning wood. *Science*, **200**, 197–200.

Woodwell, G. (1978) The carbon dioxide question. *Scientific American*, **238**(1), 34–43.

Woodwell, G. M., Whittaker, R. H., Reiners, W. A., Likens, G. E., Delwiche, C. C., and Botkin, D. B. (1978) The biota and the world carbon budget. *Science*, **199**, 141–146.

SECTION III

Soils

The total amount of carbon held in the soils of the earth is substantially greater than in either the vegetation or the atmosphere. The authors of the two papers that follow have attempted independent appraisals of the amount of carbon held in soils globally, the releases over the past century due to disturbance by man, and the current rate of release to the atmosphere. Their estimates are close for the contemporary inventory, about 1500×10^{15} g, but their estimates of the rate of release are substantially different. Buringh suggests a most probable global release of 4.6×10^{15} g while Schlesinger's estimate is about one-sixth of Buringh's, 0.8×10^{15} g.

The difference in the estimates is an indication of the need for additional studies. The most conspicuous need is for a better appraisal of the areas affected by human disturbance. We know that the major transition of importance in releasing carbon as CO_2 from soils to the atmosphere is the replacement of forest by agriculture or grazing (Woodwell *et al.*, 1978, 1983; Moore *et al.*, 1981; Houghton *et al.*, 1983). The data available at present for estimating this transition are at best uncertain (Persson, 1974; Woodwell *et al.*, 1983) because of the difficulties in determining areas on the ground and because of carelessness, confusion or bias in reporting.

There are further uncertainties. One is the rapid development of impoverished lands, especially in the tropics, where agriculture must be abandoned and where the vegetation remains sparse, diminutive or lacking. These lands are usually not tabulated as a part of agricultural or other land-use statistics. Yet they are recognized increasingly as a significant, new class of land.

The problems of measurement set forth by Buringh and by Schlesinger can be resolved in large part by remote sensing. The changes in forest soils follow major changes in the vegetation. These latter changes can be measured directly by remote sensing from aircraft and, occasionally, by satellite imagery. And so, while direct measurements of changes in soils are not usually possible using satellite imagery, indirect appraisals from direct measurements of changes in major types of vegetation are clearly possible. The papers on soils are included here to emphasize the importance of a systematic approach to improvement in the data on soils for estimation of the total flux of carbon to the atmosphere.

REFERENCES

Houghton, R. A., Hobbie, J. E., Melillo, J. M., Moore, B., Peterson, B. J., Shaver, G. R., and Woodwell, G. M. (1983) Changes in the carbon content of terrestrial biota and soils between 1860 and 1980: a net release of CO_2 to the atmosphere. *Ecol. Monogr.* (In press.)

Moore, B., Boone, R. D., Hobbie, J. E., Houghton, R. A., Melillo, J. M., Peterson, B. J., Shaver, G. R., Vorosmarty, C. J., and Woodwell, G. M. (1981) A simple model for analysis of the role of terrestrial ecosystems in the global carbon budget. In: Bolin, B. (ed), *Modelling the Global Carbon Cycle, SCOPE 16*. John Wiley and Sons, New York.

Persson, R. (1974) World forest resources: review of the world's forest resources in the early 1970's. *Research Notes Nr. 17*, Royal College of Forestry Survey, Stockholm, Sweden.

Woodwell, G. M., Hobbie, J. E., Houghton, R. A., Melillo, J. M., Moore, B., Park, A. B., Peterson, P. J., and Shaver, G. R. (1983) Measurement of changes in the vegetation of the earth by satellite imagery. This volume.

Woodwell, G. M., Whittaker, R. H., Reiners, W. A., Likens, G. E., Delwiche, C. C., and Botkin, D. B. (1978) The biota and the world carbon budget. *Science*, **199**, 141–146.

CHAPTER 3

Organic Carbon in Soils of the World

P. BURINGH

Agricultural University of The Netherlands, Wageningen, The Netherlands

ABSTRACT

The loss of organic matter from soil is mainly the result of the clearing of forests for grassland or cropland. On the basis of the assumptions outlined here, the annual loss of organic carbon from the world's soils is between 2.5×10^{15} g and 7.4×10^{15} g, with 4.6×10^{15} g being considered a realistic estimate. These amounts are 0.2, 0.5 and 0.3 per cent of the total organic carbon (1477×10^{15} g) currently estimated to exist in the world's soils. Since the total organic carbon in soil in prehistoric times has been estimated as 2014×10^{15} g, the loss since then has been 537×10^{15} g, or 27 per cent of the amount present prior to the spread of civilization in the last two millennia.

3.1 INTRODUCTION

The loss of organic matter from forest soils following disturbance is an important source of CO_2 for the atmosphere. Bolin (1977) and Schlesinger (1977, 1983) have estimated the net loss of organic carbon from the world's soil. Bolin states that if it is assumed that from 25 to 50 per cent of the presently cultivated land has been converted from forest land since the early nineteenth century, the release of organic carbon from the soil to the atmosphere during the last two centuries can be estimated at 10 to 40×10^{15} g, with an annual loss at 0.1 to 0.5×10^{15} g. Kovda (1974) has estimated the total humus in the earth's soil at 2400×10^{15} g, equivalent to approximately 1400×10^{15} g of carbon. Bohn (1976) stated that earlier in this century the organic carbon in the world's soil was estimated at 710×10^{15} g. This estimate was based on the carbon content of nine North American soils as shown in a 1915 soil textbook. Using the FAO/UNESCO soil map (1978) for South and North America, and the soil map of Ganzen and Hadrich (1965) for the other continents, Bohn estimated that there are $3000 \pm 500 \times 10^{15}$ g of organic carbon in the world's soil. Although he did not try to estimate the amount of CO_2 released to the atmosphere, Bohn (1976, p. 469) says that 'The decay of soil organic matter is one of the largest CO_2 inputs to the atmosphere'.

A few authors have tried to calculate the oxidation losses of organic matter in soils. Greenland and Nye (1959) and Nye and Greenland (1960) studied the effect of shifting cultivation on some West African soils. An excellent review has been given by Young (1976). More recently models for the decomposition of organic matter in the soil have been developed (Rayner, 1978). Unfortunately the approach used by these authors cannot be adopted for a world-wide study of soils because information is available only for a limited number of soils.

Schlesinger (1977, 1983) has calculated the mean carbon content in 11 ecosystem types and multiplied the mean by the amount of land included in these ecosystems. He offered a preliminary estimate that the earth's total soil carbon is 1515×10^{15} g. The annual release of soil carbon by his estimate is about 0.8×10^{15} g. This figure is based on the assumption that the annual conversion of forest to cultivated land is 15×10^6 ha and that the average carbon content of 131 t/ha in forest soil drops to 78.6 t/ha after conversion to cultivation, a decline of 40 per cent.

3.2   PROCEDURE

Any attempt at this time to compute the net annual release of CO_2 from the soil humus of all the soils of the world can only be a rough approximation. Detailed soil studies have only been made of a small part of the earth's land area, and 80 per cent of all soils have not been studied at all (Dudal, 1978). Most analyses of soils are incomplete because they have been carried out for agricultural rather than more general purposes.

Although organic matter is often present in the soil to a depth of 1 or 1.5 m, most is in a surface layer of from 1 to 20 cm. Carbon also exists in the mineral part of soils or in the soil solution, mainly as carbonates and bicarbonates of calcium, magnesium and sodium. Carbonates are ignored in this paper because the amount of CO_2 released to the atmosphere from mineral sources is small. The organic matter (litter) lying on top of the soil surface is also ignored because it is considered to be part of the biomass. The organic matter in soil consists of:

(a) living plant roots,
(b) dead but little-altered plant remains,
(c) partly decomposed plant remains,
(d) colloidal organic matter, being the humus proper—often some 60 to 70 per cent of the total organic matter in soils (Schnitzer and Khan, 1978),
(e) living microorganisms (bacteria, fungi, protozoans, etc.) and macroorganisms (worms, ants, termites, etc.),
(f) inactive or inert organic matter (coal, burned vegetation or ash fertilizer).

There are various methods of soil analysis to determine organic matter content (Jackson, 1958). In most analyses the soil material is passed through a two mm sieve after macroscopic plant remains (mainly roots) have been removed. Organic matter content usually is one to five per cent of the dry weight in the surface soil and decreases with depth. The carbon content is generally about 58 per cent of organic matter content (0.58 is the van Bemmelen factor), although in the tropics it is often 45 to 55 per cent. (The van Bemmelen factor is used in this paper.)

The following example gives an idea of the quantity of organic matter in one hectare. A soil with a bulk density of 1.5, and a carbon content of 3 per cent in the 0 to 25 cm layer, 1 per cent in the 25 to 50 cm layer, 0.3 per cent in the 50 to 75 cm layer and 0.1 per cent in the 75 to 100 cm layer, contains 165 t C/ha; the 0 to 25 cm layer contains 113 t/ha, or 68 per cent of the total. In a true chernozem (black earth, or mollisol) the total soil carbon is more than 200 t/ha and the surface layer of 0–25 cm contains only 25 per cent of the carbon because the humus layer is very deep.

The organic matter content in soil depends on soil conditions, present and recent vegetation cover, topography, hydrological conditions, elevation and farm management practices. Soil conditions in turn are most influenced by the soil moisture and temperature regimes, although the biological and mineralogical regimes are also important. For example, soil derived from rocks that are basic (as opposed to felsic) generally contains more organic matter than soil from felsic rocks. Clay content and type also affect organic matter content. Furthermore, the oxidation of organic matter is more rapid in calcareous soil than in non-calcareous soil. Various attempts have been made to correlate the organic matter content of soils in a specific region or country with some of these factors. Young (1976) has provided a summary of processes that affect organic matter content.

Under natural conditions the content of organic matter in soil is constant; the rate of decomposition is equal to the rate of supply of organic matter from plants. The equilibrium is disturbed when forests are cleared and the land is used for agriculture. There is also a decline in organic matter when grassland in the tropics and subtropics is transformed into cropland, or when savannahs are burned. The decline is rapid in the first few years after deforestation and gradually slows over the next 10 to 50 years. Organic matter is also lost through misuse or deterioration of land (soil erosion, salinization, alkalization and soil degradation), and because of the increasing non-agricultural use of land (urbanization and highway construction).

On the other hand, there may be an increase in organic matter when good farm management is practised and organic manure and compost are used, when arid land is irrigated, or where agricultural land is reforested. Histosols (peat soils) contain a considerable amount of carbon.

In the following pages the average soilcarbon content in various soils have been examined in relationship to types of land use. With these data the net loss of soilcarbon and the release of CO_2 to the atmosphere have been estimated as a result of changes in land use, even though precise data on the annual changes in land use are not available. Since land use in prehistoric times is also known within broad limits, it is also possible to calculate the approximate loss of soil carbon since that time as well.

A large amount of data on soil conditions and land use has been studied to establish 37 standard soils with a standard soilcarbon content. The procedure is described in the following sections.

3.3 MAJOR SOILS AND THEIR POTENTIALITIES

Initially, an attempt was made to use the 106 soil units of the FAO/UNESCO soil map as a basis for this investigation. Only a small amount of analytical data on humus content is available, however, and it was often difficult to correlate the soils described in various soil reports with the units of the soil map. Moreover, the soil units of the soil map do not indicate soil moisture and soil temperature regimes. A statistical approach based on the 10 soil orders of the new US soil classification system (Soil Survey Staff, 1975) has been adopted. Buol *et al.* (1973), and Buringh (1979) give short descriptions of the soils belonging to each soil order. The approximate extent of each soil order, and the areas that are potentially arable, non-arable but grazeable, and neither arable nor grazeable, are known (see Table 3.1). (One group, 'Mountain soils', is not subdivided into soil orders on the table.)

Table 3.1   World land area in different soil orders, in millions of hectares (10^6 ha)

Order	Potentially arable	Non-arable but grazeable	Non-arable and non-grazeable	Total	Per cent
Alfisols	640	690	400	1 730	13.1
Aridisols	80	250	2 150	2 480	18.8
Entisols	150	290	650	1 090	8.2
Histosols	1	20	100	120	0.9
Inceptisols	230	230	710	1 170	8.9
Mollisols	630	340	160	1 130	8.6
Oxosols	650	350	120	1 120	8.5
Spodosols	100	210	150	560	4.3
Ultisols	270	330	130	730	5.6
Vertisols	140	60	30	230	1.8
Mountain soils	230	910	1 670	2 810	21.3
Total	3 120	3 680	6 370	13 170	100.0
Per cent	23.7	27.9	48.4	100.0	

SOURCE: Soil Geography Unit, Soil Conservation Service, US Dept. of Agr., Washington, 1973.

Table 3.2 Major land-use types by soil orders in millions of hectares (10^6 ha)

Order	Crop land	Grass land	Forest land	Other land	Total
Alfisols	290	300	800	340	1 730
Aridisols	40	260	0	2 180	2 480
Entisols	80	200	550	260	1 090
Histosols	0	0	100	20	120
Inceptisols	150	200	650	170	1 170
Mollisols	290	680	0	160	1 130
Oxisols	300	300	450	70	1 120
Spodosols	50	150	250	110	560
Ultisols	130	150	400	50	730
Vertisols	60	100	50	20	230
Mountain soils	110	700	800	1 200	2 810
Total	1 500	3 040	4 050	4 580	13 170
Per cent	11.4	23.1	30.1	34.8	100

The potentially arable land is almost 24 per cent of the earth's total land area not covered by ice ($13\,170 \times 10^6$ ha). Similar figures have been published by the President's Science Advisory Committee (1967), Simonson (1967), Kovda (1974) and Buringh *et al.* (1975). Aubert and Tavernier (1972) and Sanchez (1976) have presented figures for soil orders in the tropics. The non-arable and non-grazeable land in Table 3.1 is mainly desert, tundra and high mountains. This classification also includes lithosols and very shallow soils.

3.3.1 Major Soils and their General Land-use Types

Four general land-use types are used in this paper. Table 3.2 shows the estimated amount of acreage of each type, divided into soil orders, in round figures. The totals are in accordance with the (1975) FAO statistical data. The total amount of potentially arable land is 3120×10^6 ha while presently arable land is 1500×10^6 ha, or 11 per cent of the total area (see Table 3.2). Total grassland area is 3040×10^6 ha, while total forest area is 4050×10^6 ha. There is almost no grassland on histosols, and no forest on aridisols or mollisols. Most of the potentially arable land not yet cultivated is grassland, and the greater part of the grazeable land not yet used is now savannah or forest. A large part of mountain soil is barren land. The soils of humid regions, such as oxisols, alfisols, utisols and most of the inceptisols and entisols, are forested if they are not used for crops or grazing.

3.3.2   Organic Carbon in Major Soils

Descriptions and analyses of more than 400 types of soil were examined for this paper. These studies show that there is great variability in total carbon content in the various types. This variability becomes much smaller when the data are grouped according to soil conditions, particularly when taking into account soil productivity, soil temperature and soil water regimes. The following observations can be made:

(a)   The analytical methods used to determine soil carbon often differ;
(b)   Soil carbon is not equal to all of the organic matter in a soil, since living plant roots and partly decomposed plant remains are excluded from the analyses;
(c)   Real soil humus is the colloidal organic matter in a soil;
(d)   Organic material lying on top of the soil (the A_0 or O-horizon, consisting of dead leaves and litter) is not included in the calculation of organic carbon;
(e)   All data are expressed in terms of organic carbon, which means that data on organic content have been multiplied by the factor 0.58;
(f)   For soils in which soil carbon has been determined to a depth of 1 m or more, soilcarbon content is less than 0.2 per cent below 1 m, and less than 0.1 per cent below 1.5 m. (Some andepts and humods that cover small areas are exceptions);
(g)   Soil surface layers (0.20 cm depth) seldom contain more than 5 per cent soil carbon;
(h)   The maximum soilcarbon content was 801 t/ha in an hydromorphic volcanic soil;
(i)   The minimum soilcarbon content was less than 10 t/ha in a desert soil;
(j)   More than half of all the soils studied had a soilcarbon content of less than 150 t/ha;
(k)   In most soil orders there is a relatively high soilcarbon content in humid climates, and a relatively low soilcarbon content in hot and dry climates;
(l)   Relatively little information is available on soil carbon in forest soils, particularly soils in virgin forests; it is estimated that a secondary forest soil contains 75 per cent of the soilcarbon content of a virgin forest (Nye and Greenland, 1960).

Table 3.3 shows the average organic carbon content of the standard soil orders of the world, but the data in Table 3.3 are not exact averages applicable to any particular country. They are based mainly on differences between the figures for various land-use types. No figures for aridisols and mollisols are given for forestlands, because aridisols are soils usually found in deserts and mollisols are usually found in prairies. Histosols represent the typical organic or peat soils of coastal swamps, which generally have not been reclaimed

Table 3.3 Average organic carbon content of representative soils in relation to major land use (soil carbon in tons per hectare)

Order	Crop land	Grass land	Virgin forest	Secondary forest
Alfisols	80	100	270	200
Aridisols	20	40	—	—
Entisols	60	110	230	170
Histosols	—	—	375	—
Inceptisols	110	140	270	200
Mollisols	130	160	—	—
Oxisols	100	150	240	180
Spodosols	80	90	130	100
Ultisols	80	110	240	180
Vertisols	70	90	190	140
Mountain soils	100	100	200	150

except for some small areas in Western Europe and North America. The figure for histosols (375 t/ha) is based on the assumption of a soil depth of 33 cm (aerated layer), a bulk density of 0.25 and a carbon content of 50 per cent.

3.3.3 Results of Basic Calculations

A calculation has been made of the total soil carbon of each soil order according to land use. Table 3.4 shows the calculation for alfisols. Here, and in all the other calculations, forest area is divided into virgin forest (50 per cent) and secondary forest (50 per cent). Moreover, it is assumed that 'other land' has approximately half of the amount of soil carbon attributed to cropland of the soil order concerned. Similar calculations are made for the other soil orders and for mountain soils.

The results of the calculations are given in Table 3.5. In addition, the percentage of the contribution of each soil order and each land-use type are given, and at the bottom the average soilcarbon content of each land-use type is presented. The total soilcarbon content of all land is 1477×10^{15} g. Kovda (1974) mentions 2400×10^{15} g of humus, which amounts to 1392×10^{15} g of organic carbon. The result of the calculation by Schlesinger (1983) was 1515×10^{15} g, while Woodwell (1978) reports various estimates, ranging from 1000 to 3000×10^{15} g. On the basis of the data in Tables 3.2, 3.3 and 3.5, it is calculated that the average loss of soil carbon after conversion of forest to cropland is 48 per cent, to grassland 28 per cent, and to agricultural land (mixed grassland and cropland) 35 per cent. (The reliability of these results is discussed in a later section.)

Table 3.4   Calculation of soil carbon in alfisols according to major land-use types

Land use (present)	Area (10^6 ha)	Soil carbon in t/ha	Total soil carbon (10^9 t)
Crops	290	80	23.2
Grass	300	100	30.0
Virgin forest	400	270	108.0
Secondary forest	400	200	80.0
Other land	340	40	13.6
Total	1 730		254.8
Land use (prehistoric)			
Virgin forest	1 390	270	375.3
Other land	340	40	13.6
Total	1 730		388.9

Table 3.5   Total soil carbon in various soil orders according to type of land, in gigatons (10^9 t $= 10^{15}$ g)

Order	Crop land	Grass land	Virgin forest	Secondary forest	Other land	Total	Per cent
Alfisols	23.2	30.0	108.0	80.0	13.6	254.8	17.3
Aridisols	0.8	10.4	—	—	21.8	33.0	2.2
Entisols	4.8	22.0	63.3	46.7	7.8	144.6	9.8
Histosols	—	—	18.8	18.7	4.0	41.5	2.8
Inceptisols	16.5	28.0	87.7	65.0	9.4	206.6	14.0
Mollisols	37.7	108.8	—	—	10.4	156.9	10.6
Oxisols	30.0	45.0	54.0	40.5	3.5	173.0	11.7
Spodosols	4.0	13.5	16.3	12.5	4.4	50.7	3.4
Ultisols	10.4	16.5	48.0	36.0	2.0	112.9	7.7
Vertisols	4.2	9.0	4.8	3.5	0.7	22.2	1.5
Mountain soils	11.0	70.0	80.0	60.0	60.0	281.0	19.0
Total	142.6	353.2	480.9	362.9	137.6	1 477.2	100.0
Per cent	9.6	23.9	32.5	25.5	9.3	100.8	
Area (mha)	1 500	3 040	2 025	2 025	4 580	13 170	
Average soil carbon (t/ha)	95	116	237	179	30	112	
Average soil carbon in all cropland and grassland:							109 t/ha
Average soil carbon in all forestland:							208 t/ha

3.4 THE DECLINE OF ORGANIC CARBON IN WORLD SOILS SINCE PREHISTORIC TIMES

It is possible to calculate the organic carbon content of soils in prehistoric times, and consequently the decline in organic carbon content since then, because certain characteristics of the subsoil horizons in soil profiles indicate the original type of vegetation. In prehistoric times there were at least 8590×10^6 ha of forest and natural grassland and 4580×10^6 ha of other land (see Table 3.2). Some 1400×10^6 ha of land were original grassland (mollisols, aridisols and half of the vertisols). The lower part of Table 3.4 shows how the total soil carbon in prehistoric times is calculated. All alfisols present on cropland and grassland once were soils in virgin forests. It is assumed that there has been no change in the carbon content of 'other soils'. The calculation of soil carbon in aridisols, mollisols and vertisols (50 per cent grassland) is based on virgin grassland.

The results of these calculations are presented in Table 3.6, which shows that the total soil carbon in prehistoric times was 2014×10^{15} g, or 537×10^{15} g (26.7 per cent) more than at present. The decline in soil carbon for each order is also shown in Table 3.6.

The average losses of soil carbon have been calculated (see Table 3.5) for the four principal land-use types, taking into account the main soil conditions of the soil orders. It is assumed that the change in land use is distributed over these soil orders proportionally. Another assumption is that any change in the organic matter content of soils that is caused by a change in land use takes place within one year. As mentioned before, it takes at least 10 to 30 years

Table 3.6 Total soil carbon in soil orders during prehistoric times and loss since then

Order	In prehistoric times		Soilcarbon decline		Per cent lost
Soil carbon	$(10^9$ t)	(per cent)	$(10^9$ t)	(per cent)	
Alfisols	388.9	19.3	134.1	25.0	34.5
Aridisols	33.8	1.7	0.8	0.1	2.4
Entisols	198.7	9.8	54.1	10.1	27.2
Histosols	41.5	2.1	—	—	—
Inceptisols	279.4	13.9	72.8	13.6	26.1
Mollisols	165.6	8.2	8.7	1.6	5.3
Oxisols	255.5	12.7	82.5	15.4	32.3
Spodosols	62.9	3.1	12.2	2.3	19.4
Ultisols	165.2	8.2	52.3	9.7	31.7
Vertisols	40.6	2.0	18.4	3.4	45.3
Mountain soils	382.0	19.0	101.0	18.8	26.4
Total	2014.1	100.0	536.9	100.0	26.7

before a new soil equilibrium is reached, but since the changes that occur in land use each year are similar, the loss is calculated as if it took place in one year.

The main problem is that global data on changing land use are scarce and unreliable, or at least suspect. Figures that seem to be accurate are often copied from earlier authors who have made rough estimates.

Three estimates of the total annual loss of carbon to the atmosphere are given in the following sections. The first, called realistic, is based on data considered realistic by the author of this paper. The low and high estimates are also based on low and high figures given in the literature.

Table 3.7 shows changes in land use for the period 1964–1974 as given by FAO (1975). In this period irrigated area increased from 194 to 226×10^6 ha, or 3.2×10^6 ha each year. Dudal (1978) mentions an increase in cropland of

Table 3.7   Change in world land use in the period 1964–1974 ($\times 10^6$ ha)

Year	Arable land	Grass land	Forest land	Other land	Total land
1964	1 412	2 045	4 062	4 866	13 385
1974	1 507	2 045	4 053	4 787	13 392
	+95	0	−9	−79	+7

SOURCE: FAO (1975) *Production Yearbook.*

between 110 and 135×10^6 ha, or approximately 6×10^6 ha each year, for the period 1957–1977.

Unfortunately, it is not known how much forest is felled, destroyed or transformed into cropland or grassland. Moreover, Table 3.7 does not indicate how much land is misused or used for non-agricultural purposes. Since the amount of grassland has not changed, it is evident that new cropland has been reclaimed from forest or from forest by way of grassland. 'Other land' seems to be reforested. Data on forest areas are especially poor (FAO, 1967), although it is well-known that there is a heavy demand for wood, that forest areas in North America and Europe are increasing, and that large areas of forest in the tropics are being felled and not reforested. The total forest area in the tropics (UNESCO/UNEP/FAO, 1978) is 1915×10^6 ha, of which 1100 are closed forest and 815 open woodlands and shrublands. Approximately 30 per cent of the world's forest area is not exploitable (Wolterson, 1977) because it is inaccessible or needed for protective and regulative functions.

3.4.1 Non-agricultural Land Use

Land is needed every year for new settlements, industries, road and highway construction, recreation, cemeteries and other uses. In Japan the amount of arable and grazing land taken annually for these purposes is 1.1 per cent of the total arable land, in Canada 0.4 to 0.9 per cent, in the Netherlands 0.6 per cent, in Poland 0.5 per cent, in the Unites States 0.5 per cent. In Great Britain, 0.7 per cent of the arable land is lost each year.

During the UN Food Conference in 1974 it was indicated that some 4 to 5×10^6 ha are used for new settlements each year. Another 6 to 7×10^6 ha are taken for other non-agricultural uses. The conference also estimated that some 150×10^6 ha, or 13 per year, would be needed for settlements. In the United States the annual loss of land to non-agricultural use is 1×10^6 ha (Pimentel *et al.*, 1976). At present, 73×10^6 ha of land in the United States are covered by settlements (Kovda, 1977). This author also estimates that by the year 2000 some 25×10^6 ha of farmland will be lost every year. A realistic estimate of the loss of cropland and grassland to non-agricultural use is 0.4 per cent of 4540×10^6 ha, or 18×10^6 ha.

New agricultural land has to be converted from forest in order to compensate for the land lost to non-agricultural uses. Assuming that intensification of agricultural production compensates for the loss of 5×10^6 ha, this means that 13×10^6 ha has to be reclaimed. The conversion of one ha of forest to cropland or grassland releases $208 - 109$ t/ha of soil carbon (Table 3.5). Therefore, the conversion of 13×10^6 ha releases 1287×10^{15} g of soil carbon to the atmosphere.

A low estimate is based on a change to non-agricultural uses of 0.3 per cent of all cropland and grassland, for a total loss of 14×10^6 ha. If it is assumed that intensified agriculture reduces the amount of new agricultural land needed to 10×10^6 ha, the release of soil carbon will amount to 990×10^{15} g.

A high estimate, based on a figure presented by Kovda (1977), is a loss of 25×10^6 ha. Assuming that intensification of agricultural purposes reduces this to 20×10^6 ha, the result would be the release of 1980×10^{15} g of soil carbon to the atmosphere.

These low, realistic, and high estimates are shown in tabular form in Table 3.8.

3.4.1.1 Land Deterioration

FAO (1978) estimated the annual deterioration of agricultural land at 6 to 12×10^6 ha, while Bivas and Bivas (1978) mentions 5 to 7×10^6 ha. Kovda (1974) said that 5 to 6×10^6 ha go out of agricultural production each year. A realistic estimate seems to be 5.3×10^6 ha, or 10 ha per minute. It is assumed here that the intensification of agricultural production compensates for the loss

Table 3.8   Low, realistic and high estimates of the soil carbon released to the atmosphere

Factor	Low estimate		Realistic estimate		High estimate	
	($\times 10^6$ t)	(per cent)	(10^6 t)	(per cent)	($\times 10^6$ t)	(per cent)
Non-agricultural use	990	39	1 287	28	1 980	27
Land deterioration	198	8	297	6	396	5
Shifting cultivation	750	30	1 250	27	1 750	24
Fuel wood, fire	250	10	750	16	1 500	20
Conversion of forest	339	13	1 017	22	1 808	24
Total	2 527	100	4 601	99	7 434	100

of 2.3×10^6 ha, and that the remaining 3×10^6 ha are regained by felling forests. The 3×10^6 ha of forest reclaimed release $208 - 109$ t/ha of soil carbon, the total release being 297×10^{12} g. A high estimate, based on the reclamation of 4×10^6 ha from forest, gives 396×10^{12} g of soil carbon released.

Kovda (1974) states that because of erosion the oceans receive 16×10^{15} g of suspended substances annually, with average organic matter of 1 to 3 per cent. If we assume a carbon content of 2 per cent, 160×10^{12} g of soil carbon is transported to the sea annually.

3.4.1.2   *Destruction of Forests Due to Shifting Cultivation*

According to several authors, up to 250 million people depend for food on shifting cultivation practices. They need at least 400 to 500×10^6 ha of forests for this purpose. If the rotation period is 15 to 20 years (it is usually shorter), these people will cut down 25 to 27×10^6 ha of forest annually. FAO estimates that 30 per cent of all forestland or 1200×10^6 ha, is involved in shifting cultivation, with 103×10^6 ha of this land situated in the Far East (FAO, 1967). Other sources estimate that the area of forest cut down every year for shifting cultivation is 8×10^6 ha in the Far East, 7 in Latin America, 7 in Africa and 2 in Oceania, for a total of 24×10^6 ha. The International Development Resource Centre (IDRC, 1976) estimated that 10×10^6 ha of forest are cut annually because of shifting cultivation, and that 100 ha of unmanaged rain forest in the tropics used for shifting cultivation can feed 30 to 50 people. Flach (1970) also came to the conclusion that in a 12-year rotation period, about two ha of forest are needed per person. But Eckholm (1976) stated that 15 ha per person were needed.

The calculations made below are based on 25×10^6 ha, 15×10^6 ha and 35×10^6 ha for realistic, low and high estimates of the area needed for shifting cultivation. This is partly virgin and partly secondary forest. During three years of cultivation the soil carbon released to the atmosphere is estimated at 50 t/ha, for a decline in organic carbon content of 1.4 per cent in the surface layer.

Calculations made with these data show that a realistic estimate of the soil carbon lost to the atmosphere annually is 1250×10^6 t, a low estimate is 750×10^6 t and a high estimate is 1750×10^6 t.

3.4.2 Destruction of Forest for Use as Fuel or Due to Forest Fire

Approximately 50 per cent of all the forest trees cut down annually in the world are used as fuel. In the tropics (Eckholm, 1976; IDRC, 1976), this figure reaches 80 per cent. More than one-third of the world population depends on wood for fuel. Most authors agree that one ton of wood is used per person per year, which is equivalent to approximately 0.5 cubic meter (m^3) of wood. In the United States in the nineteenth century, three tons of wood per capita were used for cooking and heating (Makhijiana, 1975). At present, wood use is 0.228 m^3 in the United and 0.893 m^3 in South America, where it is highest in the world (IDRC, 1976). Each year some 1000×10^6 m^3 (FAO, 1967) or some 860×10^6 m^3 (IDRC, 1976) of wood are harvested. Assuming that a hectare provides from 86 m^3 to 100 m^3, this means that approximately 10×10^6 ha of forest are cut each year.

There are few data that describe the destruction of forests by forest fire. Bourne (1978) mentions destruction by fire of 1×10^6 ha of forest in southern Para, (Brazil) in 1976. A realistic estimate would be that 5×10^6 ha of forest are destroyed by forest fire each year in the world. The loss of soil carbon during forest fires is about equal to the loss caused by shifting cultivation, 50 t/ha. Therefore, a realistic estimate of the loss caused by forest fires and by the cutting down of trees for fuel is 15×10^6 ha of forest. This would mean the release of 750×10^{12} g C to the atmosphere each year from soil.

3.4.2.1 *Conversion of Forest to Cropland and Grassland*

Large areas of forest, particularly in the tropics, are converted to grassland or cropland each year. One place where this conversion has taken place on a large scale in recent years is Amazonia (Bourne, 1978), an area of some 600×10^6 ha of virgin tropical forest, of which four-fifths is situated in Brazil. More than 10×10^6 ha of forest were cleared in Amazonia in 1975, mainly for grassland for ranching. Stellingwerf (1969) reported that between 5 and 10×10^6 ha of forest in South America were being cleared annually for agriculture. FAO (1970) reported that 8×10^6 ha of forest in Asia were being transformed every year. Hare (1978) says that 1.5 per cent, or 9×10^6 ha, of the forests in Asia are converted each year. Woodwell (1978) states that FAO data suggest that from 0.6 to 1.5 per cent—i.e., between 24 and 61 × 10⁶ ha of the world's forest—is converted annually. Kovda (1977) concluded that about 20 to 25 × 10⁶ ha of new land would be needed by the year 2000 to compensate for land loss and to meet world food requirements. FAO (1970) estimated that there was an increase of 0.7 per cent of cropland each year. According to Dudal (1978) 25 per cent of all conversion is the result of organized schemes, while 75 per cent is converted spontaneously (see also FAO, *Ceres No. 64*). At the UN World Food Conference in 1974 it was estimated that some 15 × 10⁶ ha will be converted annually in developing countries until 1985.

3.4.3 Total Forest Conversion

A realistic estimate of total forest transformation, based on the figures used in this paper, would therefore seem to be 25 × 10⁶ ha per year. Low and high estimates would be 15 and 40 × 10⁶ ha. Assuming that half of the converted forest is virgin forest and the other half secondary forest, and that the land is used for growing crops, the average decline of soil carbon would be $(208 - 95) = 113$ t/ha.

The last assumption must be made because the area of the world devoted to grassland has not increased. Thus, if forest is converted to grazing land, grassland is being converted elsewhere to cropland.

A realistic estimate of the loss of soil carbon is therefore

$$(25 - 16) \times 10^6 \text{ ha} \times 113 \text{ t/ha} = 1017 \times 10^{12} \text{ g.}$$

A low estimate would be

$$(15 - 12) \times 10^6 \text{ ha} \times 113 \text{ t/ha} = 339 \times 10^{12} \text{ g,}$$

and a high estimate would be

$$(40 - 24) \times 10^6 \text{ ha} \times 113 \text{ t/ha} = 1808 \times 10^{12} \text{ g.}$$

The 1500 × 10⁶ ha of cropland in the world are used to grow food for about four billion people. This means that 0.375 ha is available for each inhabitant of the world. But since the world's population increases about 80 million each year, 30 × 10⁶ ha of new cropland and grassland must be created each year. If we assume that about 20 per cent of the additional need for food is achieved by intensification of farming, this means that 24 × 10⁶ ha of forest has to be cut annually. A total annual clearance of 25 × 10⁶ ha of forest therefore seems to be a realistic estimation.

Although a certain amount of reforestation is occurring, particularly in North America and Europe, it takes many years before the organic carbon content of the reforested soil equals that of the original soil. Nye and Greenland (1960) found that during reforestation in the tropics the annual addition of soil carbon was 1.7 t/ha. There is also a slight increase in soil carbon when land is irrigated, when organic manure is added or when cropping is intensified, but annual addition of soil carbon due to these practices is very small in comparison to the net loss.

3.4.4 The Estimated Total Release of Soil Carbon

The results of the calculations leading to low, realistic and high estimates of the release of soil carbon to the atmosphere are summarized in Table 3.8. The total release is 2.5×10^{15} g for the low estimate, 4.6×10^{15} g for the realistic estimate and 7.4×10^{15} g for the high estimate. The destruction of forest and the conversion of forest to cropland and grassland are the main factors.

3.4.5 Histosols as a Sink for Carbon

Carbon may accumulate in swamps and peatland. The total land area covered by swamps and marshes is estimated by various authors at between 200 and 400×10^6 ha. The FAO/UNESCO soil map shows an area of 240×10^6 ha of histosols (peat soils). The increase in the peat layer is approximately 0.5 mm to 1 mm per year. This means an increase of 5 to 10 m^3 of peat per hectare. If the average bulk density is 0.25 and the carbon is 50 per cent, the total increase is 150 to 300×10^{12} g C. Thus, histosols appear to be absorbing approximately 200×10^{12} g C annually.

3.5 DISCUSSION OF THE RESULTS AND THEIR RELIABILITY

The most important results of this paper are the following:

(1) The total carbon present in the world's soils in prehistoric times, which was 2014×10^{15} g, has declined to 1477×10^{15} g. The net loss of soil carbon since prehistoric times has been 537×10^{15} g, or 27 per cent of the original amount.

(2) The total annual loss of soil carbon is computed to be 4.6×10^{15} g C (realistic), 2.5×10^{15} g (low) and 7.4×10^{15} g (high), these amounts being respectively 0.3 per cent, 0.2 per cent, and 0.5 per cent of the present quantity of soil carbon in the world.

(3) Ninety per cent or more of the annual loss of soil carbon is a consequence of the conversion of forests to grassland and cropland.

(4) The average loss of soil carbon after conversion of forest to cropland is 48 per cent, to grassland 28 per cent, and to agricultural land (mixed cropland and grassland) 35 per cent.

Three methodological aspects of the paper should also be noted:

(1) A somewhat complicated procedure had to be followed to obtain the average soilcarbon content of the various land-use types (Table 3.5) because existing differences related to soil conditions, soil climate and soil productivity had to be taken into account.
(2) Since no exact figures on the amount of land converted annually from forest are available, estimates had to be made.
(3) Since a dynamic model could not be developed because of the lack of accurate data, a bookkeeping model was used.

This study assumes that one hectare of lost agricultural land is always replaced by one hectare of land newly converted from forest. Very often, however, the new land is of lower productivity because the most productive land is already being used and because it is mostly very productive land that is being converted to non-agricultural purposes. This relationship has not been taken into account in this study.

Another factor not taken into account is the oxidation of dead roots in the soil. Also ignored is the organic matter on the surface of mineral soils, even though this matter also contributes to carbon dioxide production. Nor is any attention paid to the decline of the organic matter content of soils in relation to the decrease of soil fertility and consequently to the decrease of agricultural production.

A realistic figure for the amount of soil carbon released annually to the atmosphere is 4.6×10^{15} g. This conclusion, however, means that the recent estimates of 2×10^{15} g (Woodwell, 1978) and 0.8×10^{15} g (Schlesinger, 1983) are too low. Woodwell states that the release of carbon from the decay of soil humus is probably between 0.5 and 5×10^{15} g. The 4.6×10^{15} g calculated in this paper are almost equal to the 5×10^{15} g of organic carbon released annually by the combustion of fossil fuels. According to Schnitzer and Khan (1978), however, the decay of organic soil matter provides the largest carbon dioxide input into the atmosphere.

Another conclusion is that the total amount of soil carbon present in world soils is 1447×10^{15} g, or almost double the 827×10^{15} g of carbon in the biomass growing on world soils. The total quantity of soil carbon is in accordance with an estimate mentioned by Kovda (1974) of 1400×10^{15} g, and with Schlesinger's (1983) calculation of 1515×10^{15} g. According to McKay and Findlay (1978), the carbon content of soils is estimated to be four times the carbon content of land biota. The figure is to be compared with $3000 \pm 500 \times 10^{15}$ g presented by Bolin (1977).

The calculation on the decline of soil carbon since prehistoric times, calculated at 537×10^{15} g (or 27 per cent), is an approximation because part of the land-use type called 'other land' also probably had a forest vegetation with a higher soilcarbon content than at present. It may therefore be that the total loss of soil carbon is higher than calculated. The loss because of the conversion of virgin forest to secondary forest is already 25 per cent, and according to various authors most soils have lost 40 to 50 per cent of their organic matter content within 50 to 75 years after conversion.

3.6 ACKNOWLEDGEMENTS

I thank the United States Department of Agriculture, Soil Conservation Service for providing the data of Table 3.1 and W. M. Johnson, the Deputy Administrator of this Service, A. van Diest, P. M. Driessen, C. de Wit of the Agricultural University of Wageningen, F. Kennth Hare, Director of the Institute for Environmental Studies, University of Toronto, H. J. D. van Heemst, Centre for Agro-Biological Research in Wageningen and G. M. Woodwell, Diredtor of The Ecosystems Center, Marine Biological Laboratory, Woods Hole, USA, for commenting on a draft of this study. Moreover, I thank the participants in the SCOPE Terrestrial Carbon Conference, held in Woods Hole, Mass., USA, May 1979, for the interesting and stimulating discussion, particularly Dr Schlesinger, Dr Van Wambeke and Dr Young.

3.7 REFERENCES AND BIBLIOGRAPHY

Aubert, G., and Tavernier, R. (1972) Soil survey. *Soils of the Humid Tropics.* National Academy of Science, Washington, DC.

Bivas, M. R., and Bivas, A. K. (1978) Loss of productive soil. *International Journal of Environmental Studies,* **12**, 189–197.

Bohn, H. L. (1976) Estimate of organic carbon in world soils. *Soil Science Society of America Journal,* **40**, 468–470.

Bolin, B. (1977) Changes in land biota and their importance in the carbon cycle. *Science,* **196**, 613–615.

Bourne, R. (1978) *Assault on the Amazon.* Victor Gollancz, London.

Buol, S. W., Hole, F. D., and McCracken, R. J. (1973) *Soil Genesis and Classification.* Iowa State University Press, Iowa.

Buringh, P. (1978) Limits to the productive capacity of the biosphere. *Proceedings of the World Conference on Future Resources of Organic Raw Materials,* Toronto.

Buringh, P. (1979) *Introduction to the Study of Soils of Tropical and Subtropical Regions,* 3rd edition. Pudoc, Wageningen, The Netherlands.

Buringh, P., and van Heemst, H. D. J. (1977) *An Estimation of World Food Production Based on Labour-oriented Agriculture.* Centre for World Food Studies, Agr. Univ., Wageningen, The Netherlands.

Buringh, P., van Heemst, H. D. J., and Staring, G. (1975) *Computation of the Absolute Maximum Food Production in the World,* 1–59. Agr. Univ., Wageningen, The Netherlands.

Ceres, FAO—Review on Agriculture and Development. Rome, Italy.

Dudal, R. (1978) Land resources for agricultural development. *Proceedings of the 11th International Congress Soil Science, Edmonton.* Vol. 2, 314–340.

Elliot, W. P., and Machta, L. (eds) (1977) Study group on global environmental effects of carbon dioxide, Workshop, Miami Beach, Florida, US Dept. of Energy.

Eckholm, E. P. (1976) *Losing Ground.* W. W. Norton, New York.
FAO/UNESCO (1971–1978) *Soil Map of the World, 1:5 000 000.* Rome, Paris.
FAO (1967) Wood: World trends and prospects. *FFHC Study 16.* Rome, Italy.
FAO (1970) Five keys to development. Rome, Italy.
FAO (1975) *Production Yeargook, 1975,* Vol. 29. Rome, Italy.
FAO (1978) In *Ceres, No. 64.* July/Aug. Rome, Italy.
Flach, M. (1970) College wook 1970, 107–112. Wageningen. The Netherlands.
Ganzen, B., and Hadrich, F. (1965) *Atlas zur Bodenkunde.* Bibliographisches Institut, Mannheim, Germany.
Greenland, D. J. (1978) The responsibilities of soil science. *Proceedings 11th Congress of Soil Science, Edmonton.* Vol. 2, 341–358.
Greenland, D. J., and Nye, P. H. (1959) Increases in the carbon and nitrogen contents of tropical soils under natural forests. *Journal Soil Science,* **10,** 284–299.
Hare, F. K. (1978) Climate and its impact on renewable resources. *Proceedings of the World Conference on Future Sources of Organic Raw Material,* Toronto.
IDRC (1976) The tropical forest—overexploited and underused. *Report of the International Development Resource Centre.*
I.S.S.S. (International Soil Science Society) (1978) *Proceedings of the 11th International Congress of Soil Science Society, Edmonton, Canada,* 3 vols.
Jackson, M. L. (1958) *Soil Chemical Analysis.* Prentice-Hall, Englewood Cliffs, New Jersey.
Kononova, M. M. (ed) (1970) *Micro Organisms and Organic Matter of Soils.* Jerusalem. (Translated from Russian.)
Kovda, V. A. (1974) Biosphere, soils and their utilization. *10th International Congress Soil Science.* Moscow.
Kovda, V. A. (1977) Soil loss: an overview. *Agro-Ecosystems,* **3,** 205–224.
Lieth, H., and Whittaker, R. H. (eds) (1975) *Primary Production of the Biosphere.* Springer-Verlag, New York.
Makhijiani, A., and Poole, A. (1975) *Energy and Agriculture in the Third World.* Ballinger, Cambridge, Massachusetts.
McKay, G. A., and Findlay, B. F. (1978) The implications of climatic change. *Proceedings of the 11th International Congress of Soil Science, Edmonton.* Vol. 3, 398–425.
Nye, P. H., and Greenland, D. J. (1960) The soil under shifting cultivation. *Technical Comm. No. 51,* Harpenden, England.
Pimentel, D., Terhune, Dyson-Hudson, R., Rochereau, S., Samis, R., Smith, E. A., Denman, D., Reifschneider, R., and M. Shepard (1976) Land degradation: effects on food and energy resources. *Science,* **194**(4261), 149–155.
President's Science Advisory Committee (1967) *The World Food Problem,* Vol. 2. Washington, DC.
Rayner, J. H. (1978) Model for the decay of organic matter in soils. *Proceedings of the 11th International Congress on Soil Science, Edmonton.* Vol. 1, File 735.
Rosenburg, N. J. (1978) Weather and climate modifications on the large and small scale. *Proceedings of the 11th International Congress on Soil Science, Edmonton.* Vol. 3, 426–446.
Sanchez, P. A. (1976) *Properties and Management of Soils in the Tropics.* Wiley and Sons, New York.
Sanchez, P. A., and Buol, S. W. (1975) Soils of the tropics and the world food crisis. *Science,* **188,** 598–603.
Schlesinger, W. H. (1977) Carbon balance in terrestrial detritus. *Annual Review of Ecology and Systematics,* **8,** 51–81.

Schlesinger, W. H. (1983) Soil organic matter: A source of atmospheric CO_2. This volume.

Schnitzer, M., and Khan, S. U. (1978) *Soil Organic Matter*. Elsevier, Amsterdam.

Simonson, R. W. (1967) Present and potential usefulness of soil resources. *Annual Report 1978*, Int. Inst. Land Recl. and Improvement, Wageningen, The Netherlands, 7–25.

Soil Survey Staff (1960) Soil classification, a comprehensive system, 7th approximation. USDA, Washington, DC.

Soil Survey Staff (1977) Soil Taxonomy. *Agricultural Handbook No. 436*, USDA, Washington, DC.

Stellingwerf, D. A. (1969) The need for forest inventories. Lecture. I.T.C. Delft.

UNESCO/UNEP/FAO (1978) Tropical forest ecosystem. Paris. (Also in *Nature and Resources (UNESCO)*, **14**(3), 2.

Van Beuningen, C. S. (1977) Het bodemgeschiktheisclassificatie systeem in West Maleisië. *Yijdschr. Kon. Ned. Heide Mij.*, **88**, 209–219.

Wolterson, J. F. (1977) De vernieuwbare grondstof, bosbouw, houtgebruik en verbruik, toen, nu en straks. *Agr. Reeks*, Den Haag.

Woodwell, G. M. (1978) The carbon dioxide question. *Scientific American*, **238**(1), 34–43.

Young, A. (1976) *Tropical Soils and Soil Survey*. Cambridge University Press, Cambridge.

The Role of Terrestrial Vegetation in the Global Carbon Cycle:
Measurement by Remote Sensing
Edited by G. M. Woodwell
© 1984 SCOPE. Published by John Wiley & Sons Ltd

CHAPTER 4

Soil Organic Matter: a Source of Atmospheric CO_2

W. H. SCHLESINGER

Department of Botany, Duke University, Durham, North Carolina, USA

ABSTRACT

A careful analysis of the literature suggests that there is a pool of carbon held within the soils of the earth of about 1515×10^{15} g. Of this, human activities are causing a net release to the atmosphere of about 0.8×10^{15} g C annually. According to this analysis, the cumulative transfer of carbon to the atmosphere since prehistoric times may have reached 40×10^{15} g C.

4.1 INTRODUCTION

Much terrestrial plant debris accumulates as dead organic matter in soils. The total amount of carbon retained in this pool globally is large. Over the past century, human disturbances of soil layers have caused changes in the size of this pool; estimates suggest that between 10 and 40×10^{15} g C have been transferred to the atmosphere as CO_2 (Bolin, 1977). My purpose is to improve my earlier calculation of the size of the soil carbon pool (Schlesinger, 1977) and to estimate the cumulative transfer and current annual net loss of carbon to the atmosphere.

4.2 THE SOIL CARBON POOL

Previous estimates of the size of the soil carbon pool have differed by a factor of four (Table 4.1). The lowest estimate, 700×10^{15} g C, was based on the carbon content of nine types of US soils which were used to determine the world pool (Bolin, 1970). The highest estimate, 3000×10^{15} g C (Bohn, 1976), was based on an extrapolation from South American soil data to the world. Most recent values range from 1000 to 2000×10^{15} g C.

My 1977 estimate resulted from a search of the literature for data on the amount of carbon in surface organic matter and in underlying layers of soil. It was based on 82 values collected throughout the world. Unfortunately,

Table 4.1   Some estimates of the total carbon content of the soils of the world and net losses to the atmosphere. All values $\times 10^{15}$ g C

	Pool	Recent annual net loss	Cumulative historical loss
Bolin (1970, 1977)	700	0.1–0.5	10–40 [a]
Bazilevich (1974)	1 405		
Bohn (1976, 1978)	3 000	1–2	150 [a]
Baes *et al.* (1977)	1 080		
(cf. Olson *et al.*, 1978)			
Schlesinger (1977)	1 456		
Bolin *et al.* (1979)	1 672		
Ajtay *et al.* (1979)	2 205		
	1 636		
Buringh (1983)	1 477	4.6	537 [b]
Present paper (Table 4.2)	1 515	0.8	36 [a]

[a] Since mid-1800s.
[b] Since pre-history.

agronomists seldom measure the undecomposed organic matter on the surface of the soil and ecologists frequently do not measure the mineral horizons. Studies by soil scientists often contain data on the percentage of carbon in soils but lack the measurements of soil bulk density to calculate the amount of soil carbon per unit land area. Mean values for carbon in soil profiles, multiplied by the mean area of world ecosystems (Whittaker and Likens, 1973), yielded an estimate of soil carbon for the world of 1456×10^{15} g C. The preliminary estimate was based on relatively limited data and many assumptions. The values for boreal forests, which store large amounts of carbon, and for tropical grasslands, which cover enormous areas of the world, were especially inadequate.

My revised estimate, shown in Table 4.2, utilizes 35 new values in addition to the 82 used in the 1977 estimate. Many of the new values came from studies of the International Biological Program (IBP). Despite a 43 per cent increase in the number of values, the estimate of the world's soil carbon pool changed only four per cent.

Among forest ecosystems, mean values for carbon in the soil profile (forest floor + mineral soil) increase from the lowland tropics to the boreal region. Lowland tropical forest soils, however, are not greatly different from temperate forest soils in terms of total carbon content (Sanchez and Buol, 1975; Sanchez, 1976). High rates of organic matter production in the tropics are accompanied by high rates of decomposition. Better data from tropical forests (Pécrot, 1959; Edwards and Grubb, 1977; Grubb, 1977) have permitted separate consideration of lowland and montane rainforests, using the areas for each given by Olson *et al.* (1978). Montane tropical forests have large accumulations of soil organic matter (Table 4.2) due to the *Massenerhebung* effect (Grubb, 1971). Although production is lower at high elevations, large accumulations occur because the rate of decomposition is inhibited to a greater extent. Though less dramatic, a similar pattern is also seen on mountains in the temperate zone (Jenny, 1941; Whittaker *et al.*, 1968; Nakane, 1975; Reiners *et al.*, 1975; Franz, 1976; Hanawalt and Whittaker, 1976).

Low temperatures retard decomposition; thus, tundra and boreal forest soils contain large accumulations of soil carbon. In two spruce forests in northern Manitoba, for example, Tarnocai (1972) measured soil carbon contents of 31 and 127 kg C/m^2, with an average of 80 per cent of this carbon in the permafrost layers. Because turnover of the surface layers is markedly slower at high latitudes (Lang and Forman, 1978), it is not surprising that litter on the soil surface increases from one per cent of the total detritus in tropical forests to 13 per cent in boreal forests (Schlesinger, 1977). My mean value for boreal forest is higher than that of Kononova (1966, 1975), probably because it includes some values for northern forested peatlands.

Table 4.2 Distribution of the world pool of carbon in soil, arranged by ecosystem types for recent (1950–1970) conditions and for 1860

Ecosystem type	Mean soil profile carbon (kg C/m^2)	Recent		1860	
		World area (10^8 ha)	Portion of world soil carbon pool (10^{15} g C)	World area (10^8 ha)	Portion of world soil carbon pool (10^{15} g C)
Tropical forest					
Lowland	9.8	22.0	216	23.3	228
Montane	28.7	2.5	72	2.5	72
Temperate forest	13.4	12.0	161	13.0	174
Boreal forest	20.6	12.0	247	12.3	253
Woodland	6.9	8.5	59	10.2	70
Tropical grassland	4.2	15.0	63	16.3	68
Temperate grassland	18.9	9.0	170	11.8	223
Tundra and alpine	20.4	8.0	163	8.0	163
Desert scrub	5.8	18.0	104	18.0	104
Extreme desert, rock and ice	0.17	24.0	4	24.0	4
Cultivated	7.9	14.0	111	5.0	40
Swamp and marsh	72.3	2.0	145	2.1	152
Total		147	1 515	147	1 551

Soils in temperate zone grasslands, such as those in parts of the USSR, contain very large amounts of organic matter (Kononova, 1966, 1975). While a large percentage of the detritus in temperate zone forests is derived from litter and is rapidly mineralized on the forest floor, a major portion of grassland detritus is found deep within the soil as a result of the death of roots. This may partially explain why there is generally more carbon in the soils of temperate zone grasslands than in those of forests of the same region. The mean value for tropical grasslands, 4.2 kg C/m^2, is very low (Jones, 1973; Sanchez, 1976; Kadeba, 1978). The frequent fires may limit the amount of plant debris on the soil in these ecosystems.

Most of the values for the mean content of soil carbon in my updated estimate are slightly higher than in my 1977 review, reflecting the use of studies involving deeper sampling. Estimates for the mean content of soil carbon in tundra descreased as a result of new IBP data. The most marked change in the revised estimate, however, is due to the use of 7.9 kg C/m^2 as the typical value for cultivated soils. In 1977 the figure used was 12.7 kg C/m^2, a high value that was biased by the assumption that most cultivated land originally had the same carbon content as temperate zone grasslands. The new value, 7.9 kg C/m^2, is a weighted mean for cultivated soils worldwide that was derived from the estimates of Revelle and Munk (1977) for the area cultivated in each world ecosystem and from the assumption of a 40 per cent loss of soil carbon during cultivation.

The profile values in Table 4.2 suggest some areas in which Bohn's (1976) estimate seems too high. Bohn's value for chernozem soils, 40 kg C/m^2, is about twice my mean value for temperate grasslands and is considerably larger than the mean value of 25.3 kg C/m^2 based on many samples of chernozem soils in the USSR (Kononova, 1966). If Bohn's chernozem value is reduced by one half, his world estimate decreases by $100 \times 10^{15} \text{ g C}$. More seriously, Bohn applies the datum from one South American dystic histosol, 200 kg C/m^2, to much of the boreal region. My data for these regions are limited (Table 4.3), but no value approaches 200 kg C/m^2. If Bohn's value is reduced to 50 kg C/m^2, which is probably still too high, his world estimate is lowered by $760 \times 10^{15} \text{ g C}$. The values for carbon in the soil profiles of tropical forests (ferralsols $= 12 \text{ kg C/m}^2$) and tundra (regosols $= 20 \text{ kg C/m}^2$) given by Bohn are in good agreement with those in Table 4.2.

4.3 LOSSES OF SOIL CARBON

Each year organic matter enters the soil from plant debris derived from above and below ground sources. World-wide this input is estimated at $37.5 \times 10^{15} \text{ g C}$ of carbon per year (Reiners, 1973). At steady-state, decomposition should release an equal amount of carbon to the atmosphere as CO_2. In fact, root and mycorrhizal respiration as well as the decomposition of

Table 4.3 Soil carbon content of boreal forest soils

Location/Reference	Vegetation type	Sites	Sampling depths (cm)	Soil profile carbon (kg C/m^2)	
				Range	Mean
Canada					
Gross (1946)	Aspen, birch, spruce	9	61–91	8.0–34.8	17.5
Tarnocai (1972)	Spruce	2	180, 319	31.0–127	79.1
Leahey (1947)	Spruce	2	56, 99	15.1–41.0	28.1
Alaska					
Van Cleve (unpublished) in Edmonds (1974)	Spruce	1	100	—	31.8
Crocker and Dickson (1957)	Spruce	1	61	—	9.0
Drew and Shanks (1965)	Spruce	1	38	—	14.6
Van Cleve (personal communication)	Aspen, birch, spruce	5	33–80	17.5–28.4	22.3
Sweden					
Romell (1932)	Spruce	2	100	13.9–16.0	15.0
Nykvist (1971)	Spruce	1	50	—	10.7
USSR					
Kononova (1966)	Mean for forests on podzolic soils	?	100	—	5.0
Suvorov (1974)	Spruce	1	120	—	7.9
Ovchinnikov *et al.* (1973)	Spruce and cedar	3	150–230	12.8–15.3	13.9
Zaydel'man and Narokova (1975)	Coniferous forests	5	130–185	10.5–15.5	13.5
Mean for boreal forest					20.6

large amounts of root detritus (e.g., Edwards and Harris, 1977) result in the production of additional CO_2 which is not usually measured as input. The release of CO_2 from world soils has been reviewed recently (Schlesinger, 1977); the total steady-state carbon output from soils may be as much as 75×10^{15} g C of carbon per year.

Soil organic matter exists in many forms, Kononova (1972, 1975) separates the various forms of soil organic matter into incompletely decomposed organic matter and humus. Humus is in turn subdivided into strictly humus substances (e.g., humic and fulvic acids) and other products synthesized by soil microbes. Some humus substances are very stable. If my present estimate of

1515×10^{15} g C is divided by Reiners' (1973) estimate of 37.5×10^{15} g C as the annual production of detritus worldwide, the mean turnover time is 40 years. It is certain, however, that specific fractions of the soil organic matter in various ecosystems are far more or far less refractory. It would be instructive to make an estimate of how much soil carbon turns over in 1, 10, 100 and 1000+ years.

The clearing of forests and the conversion of forested land to farmland often reduce the soil carbon content through reduced production of plant detritus and through increased rates of erosion and decomposition of organic matter. The separation of soil carbon into fractions with different turnover rates would therefore assume great significance in evaluating soil carbon losses due to cultivation. If most of the world pool of soil carbon consists of relatively refractory carbon compounds which are unlikely to be disturbed by man's activities, we may be overemphasizing the role of soil disturbance in contributing to the amount of CO_2 in the atmosphere.

Agricultural studies give one indication of the fraction of soil organic matter that is easily decomposed. When soils are cultivated, the surface layers (0–30 cm) often lose 20 to 50 per cent of their carbon content. The rate of loss is greatest in the first few years of disturbance and slows thereafter, following the pattern typical of decomposing litter (Olson, 1963; Minderman, 1968). Harcombe (1977) listed seven studies of tropical forests in which the loss of carbon ranged from 4 to 54 per cent within one to three years after forest removal (cf. Ayanaba *et al.*, 1976; Juo and Lal, 1977). In Sierra Leone, Brams (1971) found a 50 per cent loss of organic matter in five years of cultivation (Figure 4.1). Similarly, Jones (1973) found that the cultivated soils of tropical savannahs have about one half the carbon concentrations found in undisturbed areas. Studies in temperate zone grasslands have found similar losses of organic matter from the surface layer—25 per cent in 43 years (Haas *et al.*, 1957), 31 to 56 per cent in 50+ years (Meints and Peterson, 1977) and 50 per cent in 60 years (Martel and Paul, 1974). When temperate forests are cleared for farmland, the losses of soil carbon also range up to 50 per cent (Giddens, 1957; Rubilin and Dolotov, 1967). Losses from cultivated organic soils (histosols) can be particularly great (Browder and Volk, 1978). Large additions of crop residues or manure can offset the apparent losses. The loss is often attenuated by depth. One study reported an increase in the soil carbon in the 100 to 200 cm portion of cultivated soils, but the increase was small in terms of total profile content (Meints and Peterson, 1977).

Since a large portion of the soil carbon is usually found above a depth of 50 cm, these agricultural studies suggest that nearly half of the carbon in a typical soil profile is relatively labile. The remainder is found in the lower soil profile and is very stable. In tropical regions, carbon in the lower profile is often fixed on allophane; organic matter fixed in tropical soils is often red or colourless, a fact which has perpetuated the notion that these soils are low in organic content

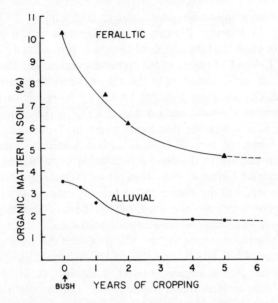

Figure 4.1   Organic matter diminution in tropical soils as a function of the number of years since clearing of native vegetation (from Brams, 1971, reproduced with permission).

(Sanchez, 1976). In other regions, crystalline clay minerals are important in complexing soil organic matter in the lower horizons (Birch and Friend, 1956; Kononova, 1966; Campbell *et al.*, 1967b; Jones 1973).

Most of the carbon lost from disturbed soils is probably lost through increased oxidation rather than erosion. Brink *et al.* (1977) report that average soil losses due to erosion in intensively farmed Wisconsin watersheds ranged from 1.2 to 1.9 kg/m^2 per year. These values are typical of all but the most careless of land clearing operations (Bormann *et al.*, 1974; Pimentel *et al.*, 1976). If we assume that these soils contain an average carbon content of two per cent, the resulting loss of carbon by erosion is rather small compared to the loss observed when a typical temperate soil is cultivated. Erosional removal of organic matter following forest cutting is probably unimportant as long as vegetation is allowed to regenerate immediately (Bormann *et al.*, 1974). Even in regrowing forests, however, the forest floor often continues to decline for a number of years after deforestation, since the young forest is not likely to produce detritus at a rate equal to the loss by decomposition. Covington (1976) measured a decline in the forest floor mass in New England for 15 years after cutting, but in subsequent years the mass increased to near steady state levels (see also Aber *et al.*, 1978).

Percolating water may transfer soil carbon to groundwater in the form of dissolved organic carbon (DOC), dissolved CO$_2$, HCO$_3^-$ and CO$_3^=$. These losses are probably minor in most natural communities (Schlesinger and Melack, 1981). In Florida, Rightmire and Hanshaw (1973) used $^{13}C/^{12}C$ isotope ratios to show that the dissolved carbon in groundwater did not reflect a change from C-3 to C-4 plants in the dominant vegetation. Thus, despite the importance of subsurface drainage in this region, subsurface carbon losses are probably slight. Groundwater losses of DOC and inorganic carbon were less than 0.3 per cent of the total carbon flux of CO_2 to the atmosphere due to decomposition in a temperate deciduous forest in Tennessee (Edwards and Harris, 1977). Similarly, various types of carbon losses in groundwater were extremely minor relative to the annual circulation of carbon in the Hubbard Brook Experimental Forest in New Hampshire (Whittaker, 1975). There has been almost no study of the change in the rate of carbon loss to groundwater when natural ecosystems are disturbed, but the data of Spalding *et al.* (1978) suggest that these losses may increase in cultivated soils.

When adjustments are made for the ^{14}C produced during atomic weapons testing, the ratios of ^{13}C to ^{12}C and ^{14}C to ^{12}C are useful in determining the age and turnover of soil organic matter (Campbell *et al.*, 1967a, 1967b; Scharpenseel, 1975; Jenkinson and Rayner, 1977; Bottner and Peyronel, 1977; O'Brien and Stout, 1978). Most of these studies have found that soil carbon consists of a small portion of relatively young material that is near the surface and turns over rapidly, and a larger amount of carbon dispersed throughout the profile with an older (500 to 2000 years) weighted mean age (Figure 4.2). The distribution of soil carbon declines exponentially with depth. O'Brien and Stout (1978) used the distribution of soil carbon enriched with bomb-derived ^{14}C to calculate a coefficient for downward diffusivity of 13 cm^2/y for carbon in a New Zealand pasture soil. Nakane and Shinozaki (1978) derived downward velocities of movement in various ecosystems from a model of soil carbon, taking into account decomposition and transport processes (see also Nakane, 1976, 1978).

The turnover time calculated by dividing the total pool of carbon by annual input is often much more rapid than indicated by the weighted mean age of carbon in the entire profile as determined by ^{14}C dating (Jenkinson and Rayner, 1977). One should remember that fractions with old weighted mean ages (e.g., humic and fulvic acids) consist of both young and old resistant carbon. Cultivation presumably causes a loss of surface organic matter and the labile portion of the more resistant fractions. As a result, the weighted mean age of the remaining material increases. Martel and Paul (1974) measured a mean residence time of 710 years for the organic matter in cultivated soils in Canadian grasslands, whereas organic matter in nearby uncultivated grasslands had a mean residence time of 250 years. O'Brien and Stout (1978) used isotope ratios to find that 16 per cent of the organic matter in the soil of a New Zealand pasture was 'very old', that is, with a minimum age of 5700 years. 'Modern' carbon was concentrated near the surface and was mostly less than 100 years old.

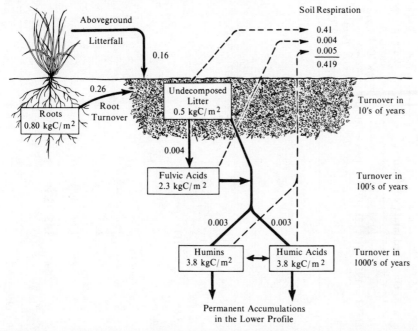

Figure 4.2 A hypothetical model for the soil carbon dynamics in the 0–20 cm layer of a chernozem grassland soil. Carbon pools (kg C/m^2) and annual transfers (kg $C/m^2/y$) are indicated. Total profile content to 20 cm is 10.4 kg C/m^2 with an overall turnover time of 25 years based on current annual input. Data are from many sources; see Schlesinger (1977)

New steadystate conditions can be achieved despite the lowered production of plant debris typical of agricultural lands, but the total amount of carbon in the soil is also lower. In conjunction with ^{14}C studies, Jenkinson and Rayner (1977) modelled the carbon dynamics in the 0–23 cm layer of a cultivated soil with an input of 100 g C/m^2 per year. Their model predicted a soil content of 2.4 kg C/m^2, about 35 per cent less than in a cultivated soil receiving 150 g C/m^2 per year and presumably less than in either area before clearing. Thus, while there are relatively few radiocarbon studies, the current evidence confirms empirical studies which show carbon losses of 20 to 50 per cent when soils are brought under long-term cultivation. I have used a 40 per cent loss factor in my estimates.

4.4 LAND CULTIVATION

Some estimates of the amount of land disturbed by human activities during the last century are shown in Table 4.4. Because many of the estimates are

Table 4.4   Estimates of the area of cultivated and disturbed land at various times in history

Source	Category	Pre-historic	Disturbed land (×10⁸ ha)							Rate of clearing (×10⁸ ha/y)	
			1860	1882	1910	1950 (1952)	1959	1966	1970	1860–1970	1950–1970
World											
Olson *et al.* (1978)	Crops	0	5						12		
	Fringe	1	3						7		
	Buildings	0	1						3		
	Total	1	9						22	0.12	
Revelle and Munk (1977)	Agricultural land	5	5			11			14	0.07	0.15
Whittaker and Likens (1973)	Cultivated					14 [a]					
Doane in Borgstrom (1969)	Tilled land			8.6		(11)					
Borgstrom from 1966 FAO Data (1969)	Tilled land							14			
North America											
Revelle and Munk (1977)	Agricultural land	0.58				2.35			2.35		
Hart (1968)	Total farmland (US only)				3.56	4.70	4.54				
	Cleared farmland (US only)				2.59	3.43	3.31				

[a] Value presumably includes all land disturbed by man; thus higher than others for 1950–1952.

based on data from the Food and Agricultural Organization (FAO), it is not surprising that there is some similarity in the values for cropland, but the estimate by Olson *et al.* (1978) strongly suggests that the total area disturbed by people is much greater than the agricultural land alone. Nevertheless, it is likely that from the time people began cultivating land to some point between 1950 and 1970, 14×10^8 ha have been converted to agriculture. This is the value used in Table 4.2.

The considerable variations in Table 4.4 are due to problems with defining cultivated land and estimating its area. One might expect some of the best data to exist for North America, yet the FAO estimates for North America (Revelle and Munk, 1977) are considerably lower than an independent estimate for the United States alone (Hart, 1968).

Revelle and Munk (1977) present a breakdown of land cultivation for each ecosystem type from 1860 to 1970, during which time they estimate that 8.5×10^8 ha were disturbed. When their data are applied to present estimates of ecosystem areas, the land area for each ecosystem in 1860 and hence the soil carbon pool at that time can be estimated (Table 4.2). On a worldwide basis, this approach suggests that approximately 36×10^{15} g C have been released from soils from 1860 to the present. Similarly, the recent rate of new cultivation (1950–1970) can be calculated as 15×10^6 ha/y (Table 4.4). If these soils contain a weighted mean of 13.1 kg C/m², which drops 40 per cent to 7.9 kg C/m² after disturbance, the present annual net release is on the order of 0.8×10^{15} g C. This estimate is higher than the range (0.1 to 0.5×10^{15} g C/yr) of potential rates of carbon release from soils suggested by Bolin (1977), but less than the value of 2×10^{15} g C/yr estimated by Woodwell *et al.* (1978). A net loss of 0.8×10^{15} g C/yr represents only a one per cent increase in the steady-state annual rate of CO_2 release from soils worldwide due to decomposition.

4.5 PROBLEMS AND APPROACHES

One of the basic tasks of soil science is to describe the physical and chemical properties of the various horizons that form a complete soil profile. The literature of soil science contains an enormous amount of these data from various parts of the world. A small subset of these data is the source of many of the profile values used in my world estimate. Further field sampling is not likely to improve our understanding of the soil carbon pool in most parts of the world, but boreal forest and tropical grassland regions are exceptions. There has been relatively little field work in these large areas, and field studies in these regions should receive high priority in the future. Soil scientists should routinely measure soil bulk density so that their measures of the percentage of carbon in the soil may be used to calculate the carbon content on an area basis with greater accuracy than is currently possible.

Rather arbitrarily, this paper's estimate of soil carbon is based on the estimates of Whittaker and Likens (1973) of the land area for world ecosystems (Table 4.2) and Revelle and Munk's (1977) estimates of disturbance (Table 4.4). As Golley (1972) has clearly indicated, there is considerable disagreement about the areal extent of world systems. Given the large variation in the estimates of disturbed land area, great improvement in our ability to estimate the release of carbon from soils could be made if the land areas were systematically studied. If we knew the extent of the world's ecosystems as identified by vegetation, and recent changes in each, we would be able to improve our estimates of the soil carbon accumulated or released in areas with different types of vegetation. Remote sensing from satellites such as LANDSAT is an obvious possibility for achieving this goal (Adrien and Baumgardner, 1977).

Alternatively, it may be possible through the use of LANDSAT to improve the mapping of soil areas at a scale appropriate for world estimates (Lewis *et al.*, 1975; Westin and Frazee, 1976; Weismiller *et al.*, 1977; Valentine, 1978). By combining data on the areal extent of soil groups with field data on soil profiles, we might be able to estimate the world pool of soil carbon with great accuracy. However, it will probably be more difficult to use this approach to estimate recent changes in the soil pool since vegetation is more visibly affected by man's activities.

The science of measuring the soil carbon content of world regions directly from satellite photographs is still in its infancy. Baumgardner *et al.* (1970) were some of the first to suggest the mapping of the organic content of surface layers of soils from aerial photographs, but subsequent studies have not always been encouraging. Evans *et al.* (1976) found that photo-tone was significantly correlated ($r = 0.57$) with the soil organic matter content of the surface layers in particular cultivated fields in Great Britain, but when the same approach was applied to a broader area of $60 \, km^2$, the correlation dropped to disappointing levels ($r = 0.31$). Cihlar and Protz (1973) also found that the correlation between photo-density values and soil properties was poor. There is apparently little possibility at the present time of estimating soil carbon content beneath areas of intact vegetation. The problem is complicated by the fact that the radiance waveband ($0.62–0.66 \, \mu m$) correlated with soil organic content (Baumgardner *et al.*, 1970) is also that of chlorophyll absorption (Adrien and Baumgardner, 1977). These studies have examined the surface layer of cultivated soils, but more have evaluated the feasibility of measuring the carbon content in an entire soil profile.

There are several ways in which soils may be serving as a sink for atmospheric CO_2. When marginal soils are irrigated and fertilized, for example, soil organic matter may increase. Hart (1968) noted that 27×10^6 ha of farmland had been abandoned in the eastern United States, and that 16.7×10^6 ha of this land had been returned to forest between 1910 and 1959. In these areas, presumably, soil carbon increases to pre-agricultural levels. In desert regions the deposition of calcium carbonate in soils may fix CO_2, although this process is likely to be minor on a global basis (Schlesinger, 1982).

Chancellor and Goss (1976) predict that 20×10^8 ha of land will be tilled by the year 2000. This would mean that 0.17×10^8 ha of land would be cleared each year for the rest of this century. On the basis of their estimate one could estimate the annual release of carbon from soils to be about 0.9×10^{15} g C per year, resulting in an additional 30×10^{15} g C being transferred to the atmosphere between now and the end of the century. Loomis (1978) presents a more apocalyptic view. Using Jenny's (1930, 1941) equations, which predict the content of soil nitrogen and organic matter in a region as a function of the mean annual temperature and a 'moisture' factor, Loomis suggests that 300 to 400×10^{15} g C might be released to the atmosphere with each $1.0\,°C$ increase in global temperature. Presumably, this release would occur predominantly in cold regions with large carbon accumulations (e.g., boreal forest and tundra). Further atmospheric warming might then occur. Because Jenny's equations have little applicability on a broad geographic basis (Schlesinger, 1977), Loomis's calculation is rather crude, but the large value obtained suggests the possible upper limit of carbon release from soils.

4.6 ACKNOWLEDGEMENTS

I thank P. Buringh, M. Drosdoff and J. T. Gray for helpful comments on an earlier version of this manuscript.

4.7 REFERENCES

Aber, J. D., Botkin, D. B., and Melillo, J. M. (1978) Predicting the effects of different harvesting regimes on forest floor dynamics in northern hardwoods. *Canadian Journal of Forest Research*, **8**, 306–315.

Adrien, P. M., and Baumgardner, M. F. (1977) Landsat, computers, and development projects. *Science*, **198**, 466–470.

Ajtay, G. L., Ketner, P., and Duvigneaud, P. (1979) Terrestrial primary production and phytomass. In: Bolin, B., Degens, E. T., Kempe, S., and Ketner, P. (eds), *SCOPE 13—The Global Carbon Cycle*, 129–181. The International Council of Scientific Unions. John Wiley and Sons, New York.

Ayanaba, A., Tuckwell, S. B., and Jenkinson, D. S. (1976) The effects of clearing and cropping on the organic reserves and biomass of tropical forest soils. *Soil Biology and Biochemistry*, **8**, 519–525.

Baes, C. F., Goeller, H. E., Olson, J. S., and Rotty, R. M. (1977) Carbon dioxide and climate: The uncontrolled experiment. *American Scientist*, **65**, 310–320.

Baumgardner, M. F., Kristof, S., Johannsen, C. J., and Zachary, A. (1970) Effects of organic matter on the multispectral properties of soils. *Proceedings of the Indiana Academy of Science*, **79**, 413–422.

Bazilevich, N. I. (1974) Geochemical work of the living substance of the earth and soil formation. *Transactions of the Tenth International Congress on Soil Science*, **6**, 17–27.

Birch, H. F., and Friend, M. T. (1956) The organic-matter and nitrogen status of east African soils. *Journal of Soil Science,* 7, 156–167.

Bohn, H. L. (1976) Estimate of organic carbon in world soils. *Soil Science Society of America Journal,* 40, 468–470.

Bohn, H. L. (1978) Organic soil carbon and CO_2. *Tellus,* 30, 472–475.

Bolin, B. (1970) The carbon cycle. *Scientific American,* 223(3), 124–132.

Bolin, B. (1977) Changes of land biota and their importance for the carbon cycle. *Science,* 196, 613–615.

Bolin, B., Degens, E. T., Duvigneaud, P., and Kempe, S. (1979) The global biogeochemical carbon cycle. In: Bolin, B., Degens, E. T., Kempe, S., and Ketner, P. (eds), *SCOPE 13—The Global Carbon Cycle,* 1–56. The International Council of Scientific Unions. John Wiley and Sons, New York.

Borgstrom, G. (1969) *Too Many.* Macmillan, London, 368 pp.

Bormann, F. H., Likens, G. E., Siccama, T. G., Pierce, R. S., and Eaton, J. S. (1974) The export of nutrients and recovery of stable conditions following deforestation at Hubbard Brook. *Ecological Monographs,* 44, 255–277.

Bottner, P., and Peyronel, A. (1977) Dynamique de la matière organique dans deux sols méditerranéens étudiée à partir de techniques de datation par le radiocarbone. *Rev. Ecol. Biol. Sol,* 14, 385–393.

Brams, E. A. (1971) Continuous cultivation of west African soils: Organic matter diminution and effects of applied lime and phosphorus. *Plant and Soil,* 35, 401–414.

Brink, R. A., Densmore, J. W., and Hill, G. A. (1977) Soil deterioration and the growing world demand for food. *Science,* 197, 625–630.

Browder, J. A., and Volk, B. G. (1978) Systems model of carbon transformations in soil subsidence. *Ecological Modelling,* 5, 269–292.

Buringh, P. (1983) Organic carbon in soils of the world. This volume.

Campbell, C. A., Paul, E. A., Rennie, D. A., and McCallum, K. J. (1967a) Factors affecting the accuracy of the carbon-dating method in soil humus studies. *Soil Science,* 104, 81–85.

Campbell, C. A., Paul, E. A., Rennie, D. A., and McCallum, K. J. (1967b) Applicability of the carbon-dating method of analysis to soil humus studies. *Soil Science,* 104, 217–224.

Chancellor, W. J., and Goss, J. R. (1976) Balancing energy and food production, 1975–2000. *Science,* 192, 213–218.

Cihlar, J., and Protz, R. (1973) Surface characteristics of mapping units related to aerial imaging of soils. *Canadian Journal of Soil Science,* 53, 249–257.

Crocker, R. L., and Dickson, B. A. (1957) Soil development on the recessional moraines of the Herbert and Mendenhall glaciers, southeastern Alaska. *Journal of Ecology,* 45, 169–185.

Covington, W. W. (1976) Secondary succession in northern hardwoods: Forest floor organic matter and nutrients and leaf fall. PhD thesis, Yale University, New Haven, Connecticut. 117 pp.

Drew, J. V., and Shanks, R. E. (1965) Landscape relationships of soils and vegetation in the forest-tundra ecotone, upper Firth River Valley, Alaska–Canada. *Ecological Monographs,* 35, 285–306.

Edmonds, R. L. (ed) (1974) An initial synthesis of results in the coniferous forest biome, 1970–1973. *US/IBP Coniferous Forest Biome Bulletin No. 7.* University of Washington, Seattle.

Edwards, N. T., and Harris, W. F. (1977) Carbon cycling in a mixed deciduous forest floor. *Ecology,* 58, 431–437.

Edwards, P. J., and Grubb, P. J. (1977) Studies of mineral cycling in a montane rain forest in New Guinea. I. The distribution of organic matter in the vegetation and soil. *Journal of Ecology*, **65**, 943–969.

Evans, R., Head, J., and Dirkzwager, M. (1976) Air photo-tones and soil properties: Implications for interpreting satellite imagery. *Remote Sensing of Environment*, **4**, 265–280.

Franz, G. (1976) Der Einflub von Niederschlag, Hohenlage und Jahresdurch-schnittstemperatur im Untersuchungsgebiet auf Humusgehalt und mikrobielle Aktivität in Bodenproben aus Nepal. *Pedobiologia*, **16**, 136–150.

Giddens, J. (1957) Rate of loss of carbon from Georgia soils. *Soil Science Society of America Proceedings*, **21**, 513–515.

Golley, F. B. (1972) Energy flux in ecosystems. In: Wiens, J. A. (ed), *Ecosystem Structure and Function*, 69–90. Oregon State University Press, Corvallis.

Gross, R. A. (1946) The composition and classification of forest floors and related soil profiles in Saskatchewan. *Scientific Agriculture*, **26**, 603–621.

Grubb, P. J. (1971) Interpretation of the 'Massenerhebung' effect on tropical mountains. *Nature*, **229**, 44–45.

Grubb, P. J. (1977) Control of forest growth and distribution on wet tropical mountains: With special reference to mineral nutrition. *Annual Review of Ecology and Systematics*, **8**, 83–107.

Haas, H. J., Evans, C. E., and Miles, E. F. (1957) Nitrogen and carbon changes in Great Plains soils as influenced by cropping and soil treatments. *U.S. Department of Agriculture Technical Bulletin 1164*. 111 pp.

Hanawalt, R. B., and Whittaker, R. H. (1976) Altitudinally coordinated patterns of soils and vegetation in the San Jacinto Mountains, California. *Soil Science*, **121**, 114–124.

Harcombe, P. A. (1977) Nutrient accumulation by vegetation during the first year of recovery of a tropical forest ecosystem. In: Cairns, J., Dickson, K. L., and Herricks, E. E. (eds), *Recovery and Restoration of Damaged Ecosystems*, 347–378. University Press of Virginia, Charlottesville.

Hart, J. F. (1968) Loss and abandonment of cleared farm land in the eastern United States. *Annals of the Association of American Geographers*, **58**, 417–440.

Jenkinson, D. S., and Rayner, J. H. (1977) The turnover of soil organic matter in some of the Rothamsted classical experiments. *Soil Science*, **123**, 298–305.

Jenny, H. (1930) A study on the influence of climate upon the nitrogen and organic matter content of the soil. *University of Missouri Agricultural Experiment Station Research Bulletin 152*. 66 pp.

Jenny, H. (1941) *Factors of Soil Formation*. McGraw-Hill, New York. 281 pp.

Jones, M. J. (1973) The organic matter content of the savanna soils of west Africa. *Journal of Soil Science*, **24**, 42–53.

Juo, A. S. R., and Lal, R. (1977) The effect of fallow and continuous cultivation on the chemical and physical properties of an alfisol in western Nigeria. *Plant and Soil*, **47**, 567–584.

Kadeba, O. (1978) Organic matter status of some savanna soils of northern Nigeria. *Soil Science*, **125**, 122–127.

Kononova, M. M. (1966) *Soil Organic Matter*, 2nd edition. Pergamon Press, Oxford. 554 pp.

Kononova, M. M. (1972) Current problems in the study of soil organic matter. *Soviet Soil Science*, 1972, 420–428.

Kononova, M. M. (1975) Humus of virgin and cultivated soils. In: Gieseking, J. E. (ed), Soil Components, Vol. 1, *Organic Components*, 475–526. Springer-Verlag, New York.

Lang, G. E., and Forman, R. T. T. (1978) Detrital dynamics in a mature oak forest: Hutcheson Memorial Forest, New Jersey. *Ecology*, **59**, 580–595.

Leahey, A. (1947) Characteristics of soils adjacent to the MacKenzie River in the Northwest Territories of Canada. *Soil Science Society of America Proceedings*, **12**, 458–461.

Lewis, D. T., Seevers, P. M., and Drew, J. V. (1975) Use of satellite imagery to delineate soil associations in the Sand Hills region of Nebraska. *Soil Science Society of America Proceedings*, **39**, 330–335.

Loomis, R. S. (1978) CO_2 and the biosphere. In: Elliott, W. P., and Machta, L. (eds), *Environmental Effects of Carbon Dioxide*. In press.

Martel, Y. A., and Paul, E. A. (1974) Effects of cultivation on the organic matter of grassland soils as determined by fractionation and radiocarbon dating. *Canadian Journal of Soil Science*, **54**, 419–426.

Meints, V. W., and Peterson, G. A. (1977) The influence of cultivation on the distribution of nitrogen in soils of the Ustoll suborder. *Soil Science*, **124**, 334–342.

Minderman, G. (1968) Addition, decomposition and accumulation of organic matter in forests. *Journal of Ecology*, **56**, 355–362.

Nakane, K. (1975) Dynamics of soil organic matter in different parts on a slope under evergreen oak forest. *Japanese Journal of Ecology*, **25**, 206–216.

Nakane, K. (1976) An empirical formulation of the vertical distribution of carbon concentration in forest soils. *Japanese Journal of Ecology*, **26**, 171–174.

Nakane, K. (1978) A mathematical model of the behavior and vertical distribution of organic carbon in forest soils. II. A revised model taking the supply of root litter into consideration. *Japanese Journal of Ecology*, **28**, 169–177.

Nakane, K., and Shinozaki, K. (1978) A mathematical model of the behavior and vertical distribution of organic carbon in forest soils. *Japanese Journal of Ecology*, **28**, 112–122.

Nykvist, N. (1971) The effect of clear felling on the distribution of biomass and nutrients. In: Rosswall, T. (ed), *Systems Analysis in Northern Coniferous Forests*, 166–178. Swedish Natural Science Research Council, Stockholm.

O'Brien, B. J., and Stout, J. D. (1978) Movement and turnover of soil organic matter as indicated by carbon isotope measurements. *Soil Biology and Biochemistry*, **10**, 309–317.

Olson, J. S. (1963) Energy storage and the balance of producers and decomposers in ecological systems. *Ecology*, **44**, 322–331.

Olson, J. S., Pfuderer, H. A., and Chan, Y. (1978) Changes in the global carbon cycle and the biosphere. *Oak Ridge National Laboratory, Environmental Sciences Division Publication No. 1050*. Oak Ridge, Tennessee. 169 pp.

Ovchinnikov, S. M., Sokolova, T. A., and Targul'yan, V. O. (1973) Clay minerals in loamy soils in the taiga and forest tundra of west Siberia. *Soviet Soil Science*, 1973, 709–722.

Pécrot, A. (1959) Quelques grand groups de sols des regions montagneuses du Kivu. *Pedologie*, **9**, 227–237.

Pimentel, D., Terhune, E. C., Dyson-Hudson, R., Rochereau, S., Samis, R., Smith, E. A., Denman, D., Reifschneider, R., and Shepard, M. (1976) Land degradation: Effects on food and energy resources. *Science*, **194**, 149–155.

Reiners, W. A. (1973) Terrestrial detritus and the carbon cycle. *Brookhaven Symposium in Biology*, **24**, 303–327.

Reiners, W. A., Marks, R. H., and Vitousek, P. M. (1975) Heavy metals in subalpine and alpine soils of New Hampshire. *Oikos*, **26**, 264–275.

Revelle, R., and Munk, W. (1977) The carbon dioxide cycle and the biosphere. In: *Energy and Climate*, 140–158. National Academy of Sciences, Washington, DC.

Rightmire, C. T., and Hanshaw, B. B. (1973) Relationship between the carbon isotope composition of soil CO_2 and dissolved carbonate species in groundwater. *Water Resources Research*, **9**, 958–967.

Romell, L. G. (1932) Mull and duff as biotic equilibria. *Soil Science*, **34**, 161–188.

Rubilin, Y. V., and Dolotov, V. A. (1967) Effect of cultivation on the amounts and composition of humus in gray forest soils. *Soviet Soil Science*, 1967, 733–738.

Sanchez, P. A. (1976) *Properties and Management of Soils in the Tropics*. Wiley, New York. 618 pp.

Sanchez, P. A., and Buol, S. W. (1975) Soils of the tropics and the world food crisis. *Science*, **188**, 598–603.

Scharpenseel, H. W. (1975) Natural radiocarbon measurements on humic substances in the light of carbon cycle estimates. In: Povoledo, D., and Golterman, H. L. (eds), *Humic Substances: Their Structure and Function in the Biosphere*, 281–292. Centre for Agricultural Publishing and Documentation, Wageningen, The Netherlands.

Schlesinger, W. H. (1977) Carbon balance in terrestrial detritus. *Annual Review of Ecology and Systematics*, **8**, 51–81.

Schlesinger, W. H. (1982) Carbon storage in the caliche of arid soils: A case study from Arizona. *Soil Science*, **133**, 247–255.

Schlesinger, W. H., and Melack, J. M. (1981) Transport of organic carbon in the world's rivers. *Tellus*, **33**, 172–187.

Spalding, R. F., Gormly, J. R., and Nash, K. G. (1978) Carbon contents and sources in ground waters of the Central Platte region in Nebraska. *Journal of Environmental Quality*, **7**, 428–434.

Suvorov, A. K. (1974) Characteristics of migration of organic and mineral substances in plowed sod-podzolic soils. *Soviet Soil Science*, **6**, 18–25.

Tarnocai, C. (1972) Some characteristics of cryic organic soils in northern Manitoba. *Canadian Journal of Soil Science*, **52**, 485–496.

Valentine, K. W. G. (1978) The 'interpreter effect' in mapping terrain in northern British Columbia using color aerial photography and LANDSAT imagery. *Canadian Journal of Soil Science*, **58**, 357–368.

Weismiller, R. A., Persinger, I. D., and Montgomery, O. L. (1977) Soil inventory from digital analysis of satellite scanner and topographic data. *Soil Science Society of America Journal*, **41**, 1166–1170.

Westin, F. C., and Frazee, C. J. (1976) Landsat data, its use in a soil survey program. *Soil Science Society of America Journal*, **40**, 81–89.

Whittaker, R. H. (1975) *Communities and Ecosystems*. 2nd edition. Macmillan, New York, 387 pp.

Whittaker, R. H., Buol, S. W., Niering, W. A., and Havens, Y. H. (1968) A soil and vegetation pattern in the Santa Catalina Mountains, Arizona. *Soil Science*, **105**, 440–450.

Whittaker, R. H., and Likens, G. E. (1973) Carbon in the biota. *Brookhaven Symposium in Biology*, **24**, 281–302.

Woodwell, G. M., Whittaker, R. H., Reiners, W. A., Likens, G. E., Delwiche, C. C., and Botkin, D. B. (1978) The biota and the world carbon budget. *Science*, **199**, 141–146.

Zaydel'man, F. R., and Narokova, R. P. (1975) Genesis of brown, podzolic, and bog-podzolic soils on coarse-textured parent materials. *Soviet Soil Science*, 1975, 11–25.

SECTION IV

Remote Sensing

EXPERIENCE AND PROSPECTS

There is now a wealth of experience in use of remote sensing to measure and interpret various attributes of the vegetation of the earth. The experience appropriate for research on the global carbon cycle embraces at once the most comprehensive and the most detailed and demanding approaches to use of satellite imagery. Aerial photography at much larger scales is the most probable basis for testing the utility and accuracy of any system based on satellite imagery.

The following papers were selected to provide a review of current uses of remote sensing in interpretation of attributes of vegetation. The reviews focus on aircraft-based imagery (Hoffer), on satellite-based imagery (Billingsley), and on the difficulties inherent in proving on the ground that which may appear obvious in remotely sensed imagery (Park). Erickson describes, in the final paper in this section, the most comprehensive programme based on satellite imagery yet executed, the Large Area Crop Inventory Experiment.

The Role of Terrestrial Vegetation in the Global Carbon Cycle:
Measurement by Remote Sensing
Edited by G. M. Woodwell
© 1984 SCOPE. Published by John Wiley & Sons Ltd

CHAPTER 5

Remote Sensing to Measure the Distribution and Structure of Vegetation

R. M. HOFFER

*Department of Forestry and Natural Resources, Purdue University,
West Lafayette, Indiana, USA*

ABSTRACT

Three types of primarily aircraft-based remote sensing systems are described with reference to global vegetation monitoring. The advantages, constraints and analysis of aerial photography, multispectral scanning systems, including LANDSAT, and radar systems are discussed. Each system provides information not obtainable from the others. Used in combination under clearly defined conditions, these systems can be applied to assess distribution and structural characteristics of the global vegetation.

5.1 INTRODUCTION

A substantial amount of study of various remote sensing systems during the past decade has shown that no single data collection platform (satellite or aircraft), remote sensing instrument system (camera, scanner or radar), or analysis technique (computer-aided analysis or manual interpretation) is adequate to meet all of the needs of various users. It appears that use of combinations of instrument systems, data collection platforms and analysis techniques will meet most needs, especially when these techniques are coupled with ancillary data. It is the purpose of this paper to examine some of the characteristics of the various instruments and techniques available for measuring the distribution and structure of the world's vegetation.

5.2 AERIAL PHOTOGRAPHY

5.2.1 Introduction

Aerial photographs have been used for many years to identify and map vegetation, to measure their area and to characterize form, size and condition of plants and plant communities. Four types of film are commonly used: black and white panchromatic, black and white infra-red, colour, and colour infra-red.

5.2.2 Black and White Films

Black and white panchromatic is the most commonly used film for aerial photography. There are many types of panchromatic film, each having its own emulsion characteristics, but as a group, panchromatic films have the best resolution of any of the film types. High resolution makes it particularly useful for such measurements as heights of trees or diameters of crowns.

Black and white panchromatic film is sensitive only to the visible wavelengths of the spectrum (0.4 to 0.7 μm), whereas black and white infra-red film is sensitive to the 0.7 to 0.9 μm portion of the spectrum as well as the visible wavelengths. Because coniferous tree species generally have less infra-red reflectance than deciduous species (Figure 5.1), black and white infra-red film is particularly useful for differentiating between these two major groups of forest cover. In some applications a Wratten 89B filter (which prevents any of the

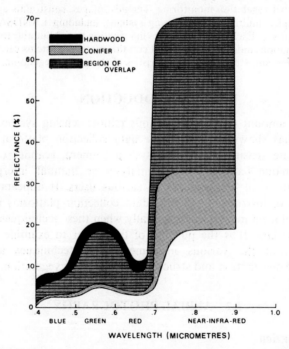

Figure 5.1   Generalized spectral reflectance characteristics of deciduous and coniferous forest species (after Murtha, 1972, reproduced by permission of the Minister of Supplies and Services, Canada)

visible wavelengths from reaching the film) is used, to provide contrast between deciduous and confierous forests. In other cases a Wratten 25 (red) filter is preferred, so that the film is sensitized by both the red visible and reflective infra-red wavelengths. The result is a photograph with less contrast. Such 'modified black and white infra-red' photos are commonly used by the US Forest Service. Because the resolution of black and white infra-red film is not as good as that of panchromatic, the latter film is preferred for mensurational purposes.

Since black and white infra-red film is sensitive to the visible as well as to the infra-red wavelengths, it is the film used for multiband photography. A considerable amount of research was conducted in the late 1960s using multiband cameras, most of which had four lenses, but some of which had as many as nine (Lowe *et al.*, 1964; Yost and Wenderoth, 1967; Colwell, 1968; Lauer, 1971; Reeves, 1975). Different filters on each lens allow only certain wavelengths to impinge on the film. Through the use of a special viewing device and appropriate filters, it is possible to combine images from the blue, green and red visible wavelengths, for example, to create a standard colour image of the scene; or the green, red and reflective infra-red wavelength bands can be combined to create a 'colour infra-red' image. Other combinations can also be defined to enhance a particular feature of interest. Although these methods provide a relatively inexpensive research tool, they have seldom been used on an operational basis.

5.2.3 Colour Films

Black and white photos are generally not as useful as colour or colour infra-red photos for identifying individual species of trees, shrubs, grasses and forbs. This is not surprising, since the human eye can distinguish far more hues and tones of colour than it can distinguish shades of grey (Heller, 1970). Colour films have three emulsion layers that are sensitive to the blue, green and red visible wavelength portions of the spectrum. Some colour films, such as Kodachrome and Ektachrome, are colour reversal films that are developed into a positive emulsion (or transparency) suitable for direct viewing. Other colour films, such as Kodacolor and Ektacolor, produce a film negative from which positive prints are made. The resolution of such prints is not as good as that of the transparencies, but prints are easier to use in the field. There is also a colour film called Aero-neg, which can be developed into either a positive or negative emulsion. Developed as a negative, this film can then produce positive prints, transparencies or diapositive plates in either black and white or colour (Smith, 1968).

5.2.4 Colour Infrared Film

Colour infra-red film (i.e., Kodak Aerochrome Infrared Film, Type 2443) has been tested and used extensively for mapping vegetation and assessing its condition. Because the film is sensitive to wavelengths to which our eyes are not sensitive, it is more difficult to interpret than regular colour film. Colour infra-red is similar to regular colour film, however, in that both types have three emulsion layers. The main difference is that the three emulsion layers of properly filtered infra-red film are sensitive to the green, red and reflective infra-red wavelengths. Since all three emulsion layers of colour infra-red are also sensitive to the blue wavelength portion of the spectrum, it is used with a yellow (or 'minus-blue') filter (usually a Wratten 12 or Wratten 15) to obtain a good image. Thus, regular colour film has a sensitivity range of about 0.4 to 0.7 μm whereas properly filtered colour infra-red film is sensitized to wavelengths between 0.5 and 0.9 μm (Kodak, 1976).

It should be pointed out that photographic films used in remote sensing are limited to the ultraviolet, visible and reflective infra-red wavelengths up to 0.9 μm. The thermal infra-red portion of the spectrum, however, extends from approximately 3 to 14 μm. Hence, *colour infra-red film is not sensitive to the thermal infra-red portion of the spectrum*, and *cannot be used* to detect thermal phenomena, such as heated water discharged from hydroelectric plants into rivers or lakes, or heat loss from buildings (Fritz, 1967). Thermal infra-red scanner systems must be used to detect energy in the thermal infra-red portion of the spectrum.

One of the major advantages of colour infra-red film is its ability to enhance subtle differences in reflectance that are barely discernible in the visible wavelengths (Reeves, 1975). Frequently, spectral differences due to variations in plant species or to stress conditions will exist but will be so subtle that they are difficult to see on regular colour film. Although spectral differences may be very small in the visible wavelengths, they may be very distinct in the near infra-red wavelengths and therefore will show up clearly on colour infra-red film. Examination by the author of a large number of colour infra-red photos and corresponding colour photos obtained at the same time, and a careful review of the literature indicate that there are relatively few cases in which spectral differences in vegetation, soils or water that are visible on colour infra-red film cannot be detected on properly exposed colour film. However, there are many cases where the spectral differences are so subtle that they would be missed if the photo interpreter relied only on regular colour film.

A second major advantage of colour infra-red film is its ability to penetrate atmospheric haze better than normal colour film. This is because atmospheric scattering of light is more pronounced in the shorter wavelengths, and the yellow filter normally used with colour infrared film prevents these strongly scattered blue wavelengths from reaching the film.

5.2.5   Interpretation of Colour and Colour Infra-red Films

Whereas the cyan emulsion layer of colour infra-red film is sensitive to the 0.7 to 0.9 μm reflective infra-red wavelengths, the other two emulsion layers are sensitive to the visible wavelength portion of the spectrum (green and red). This implies that spectral variations in either the visible or the reflective infra-red wavelengths, or a combination of both, will cause colour differences on colour infra-red film. Thus, even though different objects have different colours on colour infra-red film, it does not necessarily follow that a difference in infra-red reflectance is present. The difference in colour may be caused solely by differences in reflectance in the visible wavelengths. Frequently, however, differences in colour on the colour infra-red film are caused by spectral variations in both the visible and infra-red wavelengths. This makes colour infra-red film difficult to interpret unless something is known about the spectral characteristics of the material of interest, both in the visible and infra-red wavelengths.

A considerable amount of research has been conducted into the usefulness of colour and colour infra-red films for mapping vegetative types and conditions. Distinguishing between deciduous and coniferous trees can be done very effectively with colour infra-red film, since the higher near infra-red reflectance of the deciduous species cause them to have a much brighter red appearance than the conifers. Determining the amount of vegetative cover present in an area is also much easier with colour infra-red film, which enhances the appearance of vegetation against a soil background. Colour film is much better than black and white panchromatic film for identifying individual species of trees (Heller *et al.*, 1966) and colour infra-red has been shown to be more effective than colour film for identifying a variety of grasses, forbs and shrubs (Driscoll and Coleman, 1974). The effectiveness of colour or colour infra-red film for differentiating species is often dependent on the phenology of the vegetation as well as the scale and quality of the photos used. Because of the increased information content and interpretability of colour and colour infra-red films, efficiency of interpretation is increased significantly (Lauer, 1971). In one study, the time required to classify 50 000 acres of forest land was cut from 44 to 21 hours through the use of small-scale colour infra-red film rather than black and white panchromatic (Lauer and Benson, 1973). The advantages of regular colour film have led the US Forest Service to adopt 1:15 840 colour photography for National Forest mapping activities.

When plants are affected by stress, such as that caused by disease or insect damage, changes occur in the spectral reflectance characteristics of the foliage. Colwell (1956) reported previsual detection of wheat rust using colour infra-red film, provided the photos were obtained under certain conditions of development of the disease, illumination and film exposure. Manzer and Cooper (1967) showed that colour infra-red film can be an effective tool for detecting late blight in potatoes. There have been many other studies concerning the use of colour infra-red film, as well as other film types and remote sensing systems, to detect stress (Colwell, 1960; Heller, 1971; Bauer *et al.*, 1971; Murtha, 1972; Reeves, 1975; Aldrich, 1979). The enhancement capabilities of colour infra-red film clearly make it a very useful tool for monitoring plant diseases and insect infestations. Such conditions cause a difference in tone that makes the stressed vegetation distinguishable from the normal red tone of healthy surrounding vegetation. However, it would appear that there are very few well-documented cases of true previsual stress detection using colour infra-red film. Aldrich (1979) stated that there was no evidence to indicate that previsual stress could be detected in either coniferous or deciduous trees.

To assess the state of the art concerning the capabilities of remote sensing for assessing vegetation damage, a special symposium was sponsored by the American Society of Photogrammetry in 1978, the proceedings of which are available from ASP. The theory and use of remote sensing for vegetation damage assessment are summarized very well in papers by Murtha (1978) and Heller (1978).

5.2.6 Scale of Photography

In addition to the type of film used, the scale of aerial photography affects its utility for vegetation mapping and monitoring. Most activities involving remote sensing to map and characterize vegetative cover are concerned with floristic mapping, physiognomic mapping or stress detection and monitoring. For these purposes, medium to large-scale photos are generally needed to achieve accurate and reliable results. In the United States, foresters have traditionally utilized a scale of 1:15 840 (four inches equals one mile), while agronomists and soil scientists have preferred 1:20 000 scale photos (Colwell, 1960). Improved film resolution, cameras and aircraft capabilities now allow smaller scale photos to be used for some purposes, and 1:40 000 is currently the standard scale used by USDA in many states for crop surveys and soils mapping.

The particular scale and the type of film to be used depend on the degree of mapping detail involved and the accuracy required. For most types of quantitative measurement, as photo scale decreases so does the accuracy of measurement. For example, 1:15 840 is an effective scale for mapping forest cover types, but identification of individual trees by species and measurements of height and crown diameter to obtain volume estimates require stereo photos that have a much larger scale, such as 1:1 000 to 1:5000 (Heller *et al.*, 1966; Sayn-Wittgenstein and Aldred, 1967; Avery, 1977; Aldrich, 1979). Crown closure estimates can be obtained from somewhat smaller scale photos, such as 1:5000 to 1:15 000 (Avery, 1977). Rangeland managers require very large-scale (1: 800 to 1:1500) photos to identify individual species of shrubs and range vegetation, although smaller scale photos can be used to delineate vegetation communities and their condition. Colour and colour infra-red photos are much more effective than panchromatic or black and white infra-red photos for these purposes (Driscoll and Coleman, 1974). Table 5.1 provides a good summary of the relationships between the scale of the photography and the degree of detail that can be obtained.

Table 5.1   Utility of different scales for vegetation mapping (from Avery 1977)

Type of imagery or scale	General level of plant discrimination
Earth-satellite imagery	Separation of extensive masses of evergreen versus deciduous forests
1:25 000–1:100 000	Recognition of broad vegetative types, largely by inferential processes
1:10 000–1:25 000 ´	Direct identification of major cover types and species occurring in pure stands
1:2500–1:10 000	Identification of individual trees and large shrubs
1:500–1:2500	Identification of individual range plants and grassland types

Careful consideration should always be given to the purpose for which the photos are to be used since, as the scale increases, complete coverage of an area will require more flight-lines and also result in a much larger number of photos, thereby increasing both data collection and handling costs significantly. For example, 23 cm × 23 cm (9 in × 9 in) stereo photos for an area of 1000 km^2 would require 45 photos at a 1:40 000 scale, 177 photos at a 1:20 000 scale, 705 photos at a 1:10 000 scale and 2778 photos at a 1:5000 scale (Avery, 1977). It is apparent, therefore, that the smallest-scale photo that is adequate to provide the information required is the most economical scale to use.

5.2.7   Use of Small-scale Photography

Several studies during the past decade have helped to define many of the potentials and limitations of small-scale aerial (1:120 000) and space (1:500 000 to 1:2 400 000) photos (Draeger *et al.*, 1971; Aldrich, 1971; Lauer and Benson, 1973; Hay, 1974; Marshall and Meyer, 1977; NASA, 1978). Small scale colour infra-red photos have been shown to be effective for distinguishing forest from non-forest classes and for differentiating deciduous from coniferous forest cover. Individual forest cover types (i.e., species associations) generally could not be identified directly, but in some cases the boundaries of different cover types could be delineated, and through comparison with larger scale aerial photos or existing reference data, the type could be identified. The degree of detail that could be defined depended on the season and the characteristics of the forest and other vegetative cover types present.

Time of year has been shown to be particularly critical in agricultural applications of small scale aerial and space photographs, since crops develop rapidly and different species are often at different stages of development at any particular time. Most studies of agricultural applications have concluded that data are needed at more than one time during the growing season, and that dates when photos are needed are a function of the crop calendars for the various species (Colwell, 1960; Reeves, 1975; Bauer, 1975; NASA, 1978).

When using medium to large scale photos from aircraft altitudes, the interpreter generally utilizes many (if not all) of the commonly defined principles of photo interpretation to identify cover types. These principles include size, shape, tone and colour, texture, shadow, pattern and association (including site). However, when using the very small-scale photos obtained from spacecraft altitudes the interpreter finds himself much more dependent on tone or colour, because such characteristics as shadow and texture or the size and shape of individual trees cannot be discerned. Dependence on the colour of various cover types, and the variability of the colour as a function of time of year and geographic location, also make the interpreter much more aware of the need to be knowledgeable about the spectral reflectance characteristics of various cover types and how such spectral characteristics vary, both as a function of time and of geographic location.

One of the major advantages of spacecraft data is its synoptic view. Hence, the use of small-scale photos to cover an entire area and to delineate or stratify major cover types at a generalized level, in combination with statistically defined samples of medium and large-scale photos to identify individual cover types and their characteristics, seems to be a logical approach to obtain reliable resource information. Such an approach, often referred to as 'multistage' or 'multilevel' sampling, allows one to take advantage of the capabilities of the various scales (Langley *et al.*, 1969; Heller, 1978; Aldrich, 1979).

One obvious limitation of photographs obtained in space is the difficulty of returning them to earth. Other than the Apollo-9 and Skylab EREP projects, there have been relatively few photographic studies of earth resources from space. However, these two projects did create a great deal of interest in mapping and monitoring vegetation and other earth resources from space. Thus, the launch of LANDSAT-1 and the capability to telemeter this type of data received considerable interest from resource managers.

5.3 MULTISPECTRAL SCANNER SYSTEMS (MSS)

5.3.1 Introduction

The launching of the LANDSAT-1 (originally ERTS-1) Earth Observation Technology Satellite in 1972 greatly increased the use of multispectral scanner data. Prior to 1972, multispectral scanner systems (MSS) had been flown at aircraft altitudes, and the possibility of identifying various features of the earth's surface on the basis of spectral reflectance patterns had been shown (Lowe *et al.*, 1964; Hoffer, 1967). The first MSS capable of obtaining data throughout the optical portion of the spectrum and recording the data on tape was developed at the University of Michigan in 1966, and the first single-aperture system became available in 1971 (Hasel, 1972). Early work with MSS data indicated that the increased range of wavelengths in which data could be obtained offered significant potential, but that manual interpretation of subtle differences in reflectance or emittance among many different images was not an effective method for analysing such data (Hoffer, 1967). The concept of applying pattern recognition techniques to the analysis of multispectral scanner data was then developed and in 1967 it was demonstrated that such an approach was feasible (Landgrebe and Staff, 1967; Holter *et al.*, 1970). Since that time, many techniques for computer-aided analysis of MSS data have been developed.

5.3.2 Multispectral Scanner Systems

Multispectral scanner systems differ from photographic systems in several ways, including the optical-mechanical mechanisms for collecting data, the quantitative character of the data collected, and the range of frequencies to which the detectors are sensitive.

In multispectral scanner systems the energy reflected or emitted from a small area on the earth's surface (the resolution element or instantaneous viewing area) at a given moment is reflected from a rotating or oscillating mirror through an optical system which disperses the energy spectrally on to an array of detectors. The motion of the mirror allows the energy along a scan line (which is perpendicular to the direction of flight) to be measured, while the forward movement of the aircraft or spacecraft brings successive strips of terrain into view. The detectors, carefully selected for their sensitivity to energy in the various portions of the spectrum, and appropriately filtered, simultaneously measure the energy in the different wavelength bands. The output signal from the detectors is amplified and recorded on magnetic tape.

The quantitative format of MSS data makes it ideally suited for telemetering to earth and for processing by computer-aided analysis techniques, whereas photographic data are qualitative in format and best suited for manual interpretation. The spatial resolution of scanner data (i.e., the instantaneous viewing area on the ground) is a function of both the characteristics of the scanner and its altitude. Since the data from scanner systems generally do not have spatial resolution as good as can be obtained from photographic systems at the same altitude, small objects cannot be resolved as well. However, the spectral resolution of MSS systems can be much better (i.e., energy from much narrower wavelength bands can be accurately measured). Of perhaps even more importance is the fact that scanners can record data throughout the 0.3 to 14 μm wavelength region, but photographic systems cannot effectively record data at wavelengths longer than 0.9 μm.

5.3.3 Spectral Reflectance Characteristics of Vegetation

The data collected by multispectral scanners represent the spectral reflectance and emittance characteristics of various cover types. It has been determined that different cover types reflect and emit varying amounts of energy in a single spectral band, and that a single object reflects and emits varying amounts of energy as a function of wavelength (Gates *et al.*, 1965; Hoffer and Johannsen, 1969; Howard, 1971; Sinclair *et al.*, 1971; Hoffer, 1978). Therefore, the proper interpretation of multispectral scanner data or other remote sensor data (such as colour infra-red film) requires a knowledge of the spectral characteristics of vegetation, soil, water and other earth surface features.

Figure 5.2 is an example of the spectral reflectance characteristics of typical green vegetation. This curve shows the low reflectance due to chlorophyll absorption bands at approximately 0.45 and 0.65 μm in the visible wavelengths, the typical high reflectance in the 0.72 to 1.3 μm (near infra-red) region, and the distinct water absorption bands at approximately 1.45 and 1.95 μm in the middle infra-red wavelengths. Minor water absorption bands are also evident at about 0.96 and 1.2 μm.

In the visible and reflective infrared portions of the spectrum, the energy incident (I) upon an object that is not reflected (R) by the object must be either absorbed (A) or transmitted (T) through the object. Thus, for any particular wavelength (λ):

$$I_\lambda = R_\lambda + A_\lambda + T_\lambda$$

For turgid green vegetation, most of the energy in the visible wavelengths (below about 0.72 μm) is absorbed by chlorophyll, with less absorption and higher reflectance in the green wavelengths (about 0.55 μm) between the two chlorophyll absorption bands. Very little energy in the visible wavelengths is transmitted through a leaf, but in the near infra-red wavelengths (from about 0.72 to 1.3 μm) only very small amounts of energy are absorbed, and nearly all energy not reflected is transmitted through the leaf. In the middle infra-red

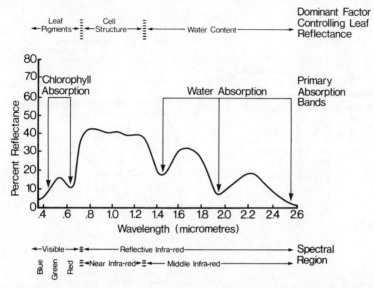

Figure 5.2 Spectral reflectance characteristics of green vegetation (after Hoffer and Johannsen, 1969)

wavelengths from 1.3 to 2.6 μm, most of the energy not absorbed by the water in the leaf is reflected, leaving much smaller amounts to be transmitted (Gates *et al.*, 1965; Hoffer and Johannsen, 1969). In these wavelengths the amount of energy absorbed is a function of the water content of the leaf, which is related to both moisture content and leaf thickness (Hoffer, 1978).

Although some vegetative cover types have significantly different spectral response patterns (as seen in Figure 5.1), many tree species and many agricultural crops have spectral response patterns that are very similar, as indicated in Figure 5.3. Although Figure 5.3 shows differences between the curves of various species, such differences are often not consistent or distinct enough to permit species identification when measurements from multispectral scanner systems at altitudes ranging from a few thousand feet to hundreds of miles are used. However, spectral differences between major cover types, such as green vegetation, dry dead vegetation, light and dark soils, clear and turbid water and snow, are significant and distinct, as indicated in Figure 5.4. It is the spectral differences that enable multispectral scanner systems and either manual or computer-aided analysis techniques to be used to identify and map various surface features and cover types. Use of MSS data from the LANDSAT satellites has been the subject of numerous investigations during the past several years.

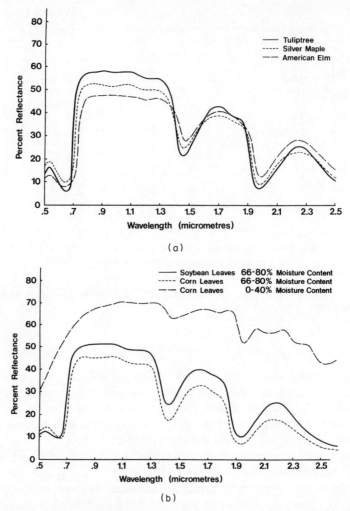

Figure 5.3   Spectral reflectance for (a) three species of trees and (b) corn and soybeans (after Hoffer and Johannsen, 1969)

5.4   THE LANDSAT MULTISPECTRAL SCANNER SYSTEM

5.4.1   Introduction

The launch of LANDSAT-1 in 1972 opened a new dimension in our capability to obtain data over most of the earth's surface at any time of year (cloud cover permitting). LANDSAT-1 was followed by LANDSAT-2 in 1975, LANDSAT-3 in 1978 and LANDSAT-4 (LANDSAT-D) in 1982. The polar orbit of the

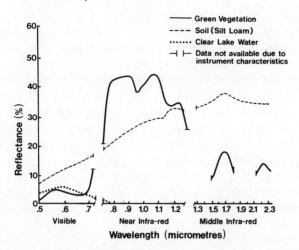

Figure 5.4   Spectral reflectance characteristics of major earth surface cover types (after Bartolucci *et al.*, 1977)

satellites allows data to be collected over virtually the entire surface of the earth every 18 days. The MSS system on the satellites collects data in four wavelength bands, including the 0.5 to 0.6 μm (green visible), 0.6 to 0.7 μm (red visible) and 0.7 to 0.9 μm and 0.8 to 1.1 μm (reflective infra-red). The data are handled in frames, each of which covers an area on the ground of 185 × 185 km (115 × 115 statute miles). A frame contains 2340 scan lines, each of which has 3236 resolution elements, resulting in a total of 7 572 240 resolution elements, or pixels, each of which represents an area of approximately 0.46 hectare (NASA, 1972a). Since the data are collected in each of the four bands, more than 30 million individual reflectance measurements are contained on the data tape representing a single frame, which is obtained in only 25 seconds. The Thematic Mapper (TM) scanner on LANDSAT-4 obtains significantly greater quantities of data because it has higher spatial resolution (i.e., 30 m) and seven spectral bands.

LANDSAT data for any portion of the earth's surface for which they exist are available through the EROS Data Center, Sioux Falls, South Dakota. The data can be obtained as a black and white image at scales of 1:1 000 000, 1:500 000 or 1:250 000, or as a 'colour infra-red composite' at one of the same scales (Rhode *et al.*, 1978). The colour infra-red composites are artificially created through appropriate combination of the green, red and one of the reflective infra-red wavelength bands, and are therefore often referred to as 'false' colour infra-red images.

5.4.2   Interpretation of LANDSAT Data

Manual interpretation of LANDSAT data has been used in many studies throughout the world to map various types of vegetation and its characteristics (NASA, 1972b, 1973a, 1973b, 1975; Heller, 1975; Aldrich, 1975, 1979; Adrien and Baumgardner, 1977; Hooley *et al.*, 1977; Hsu, 1978). In general, these studies have indicated that the colour infra-red composite at a scale of 1:250000 is most effective. At this scale and in this format, major cover types—such as deciduous forest, coniferous forest, rangeland, agricultural lands, tundra, areas of exposed soil or rock and water (both clear and turbid)—generally can be delineated and mapped with a reasonable degree of accuracy. Disturbed forest lands, such as clear-cuts, can also be reliably defined (Lee, 1975; Aldrich, 1975). However, work by Hsu (1978) and the author with LANDSAT data in Bolivia, Taiwan and Thailand indicates that it is difficult to define brushland areas, such as those that develop after clear-cutting, unless the clear-cut is relatively recent. After that, the spectral response from brushland is very similar to the response from jungle canopy.

5.4.3   Computer-Aided Analysis Techniques

Since LANDSAT data are obtained in digital form, and because such a format is well-suited to quantitative data processing, there has been considerable interest in applying computer-aided analysis techniques (CAAT) to LANDSAT data.

Two basic types of computer processing techniques can be applied: enhancement and classification. Enhancement techniques are directed at producing an image which is manually interpreted. This may involve enhancing the spectral characteristics in the data or digitally enlarging the data to allow more effective interpretation of spatial features (Kourtz and Scott, 1978). It may also involve superimposing two sets of data and enhancing the temporal differences.

Classification generally involves a series of steps including data reformatting and preprocessing, definition of training statistics, computer classification of the data, information display and tabulation, and evaluation of the results. These are discussed in more detail below.

5.4.3.1   Data Reformatting and Preprocessing

Data reformatting and preprocessing involves such activities as reformatting scanner data to place a full frame on a single data tape, digital filtering to improve data quality (signal-to-noise ratio), geometrically correcting and scaling the data to a common map base, and digital registration of multiple sets of scanner and other data. Such procedures do not involve analysis of the data. They simply allow data analysis to be carried out in a more effective manner.

5.4.3.2 Definition of Training Statistics

Analysis of MSS data involves a series of steps designed to enable the computer to identify various cover types or earth surface features. The key element is that the computer is 'trained' to recognize the particular combinations of numbers (the reflectance measurements in each of the wavelength bands) that characterize the cover types of interest. This training process involves scanner data from a limited geographic area. After a good set of training statistics has been developed, the computer is programmed to classify the reflectance values for each resolution element in the entire data set. In this way the computer can be used to map and tabulate cover types over a large geographic area at a much faster rate than is possible by using standard image interpretation techniques.

One of the major considerations in developing training statistics is the definition of the classes of material that the computer should be trained to recognize. There are two conditions which must be met by each class in an analysis of multispectral scanner data:

(1) The class must be spectrally separable from all other classes.
(2) The class must be of interest to the user or have informational value (Hoffer, 1976a).

One often finds that the classes of interest to the user cannot be spectrally separated at certain times of the year. Quite often, different species of green vegetation have very similar spectral characteristics, even though their morphological characteristics are quite different. Because a class must both be separable and have informational value, two quite different approaches are used in training the computer system.

The first approach is referred to as the 'supervised technique' and involves the use of a system of $X-Y$ coordinates to designate to the computer system the locations of known earth surface features that have informational value. For example, a certain $X-Y$ location is designated as a stand of ponderosa pine, another as a stand of aspen, and others as grassland, water, etc. This technique has been used quite effectively for agricultural mapping, but experience has shown that it is not as reliable for wildland areas, where the cover types of interest are not as spectrally homogeneous. The primary reason for this is the difficulty of defining locations that are representative of all the variations in spectral response for every cover type of interest (Hoffer and Fleming, 1978).

A second approach is the 'clustering technique' (sometimes referred to as the 'non-supervised technique'). In this approach the analyst designates the number of spectrally distinct classes into which the data to be classified should be divided. The computer is then programmed to classify the data into the designated number of classes and to print a map indicating which resolution elements belong to which spectral class. The analyst then compares this map with surface observation data (usually aerial photos) and determines which materials are represented by each of the different spectral classes on the map (e.g., spectral class one is aspen, class two is ponderosa pine, etc.). One problem with the technique is that the analyst does not know how many spectral classes are actually present. Furthermore, the classes of most interest are often rather similar spectrally, while many of the other classes may be easily separated spectrally but are of little informational value. In spite of these difficulties, much of the early work with the clustering technique indicated that it was more effective than the supervised technique for wildland or natural areas (Smedes *et al.*, 1970). With the advent of LANDSAT-1, computer-aided mapping of relatively large areas became more feasible. It was then found, however, that the amount of data and the number of spectral classes became too large to use the clustering technique effectively.

A 'multi-cluster blocks technique' was therefore developed and has proven to be extremely effective (Hoffer and Fleming, 1978). This technique involves a combination of the clustering and supervised approaches. Several small blocks of data (e.g., 40×40 pixels) are defined, each of which contains several cover types. Each data block is first clustered separately, and the spectral classes for all cluster areas are then combined. In essence, the modified cluster approach entails discovering the natural spectral groupings present in the scanner data and correlating the resultant spectral classes with the desired informational classes (cover types, vegetative conditions, etc.). Often, less than one per cent of the data involved in the final analysis are used for the training phase.

5.4.3.3　Computer Classification of MSS Data

After the training statistics are defined, the MSS data for the entire area of interest must be classified. The basis of the theory behind most of the classification algorithms is illustrated in Figure 5.5. Multispectral scanners are designed to measure the relative reflectance or emittance in designated wavelength bands, as indicated in Figure 5.5(a) by λ_1, λ_2 and λ_3. By plotting the relative reflectance (i.e., response values) of vegetation, soil and water for λ_1, λ_2 and λ_3 (as shown in Figure 5.5(b)), one sees that these cover types occupy very different locations in three-dimensional space. The classification algorithm must divide this three-dimensional space into regions that can be used to 'classify' any unknown data points or vectors. Any unknown data vector is classified into one of the spectral classes which the computer has been trained to recognize. Any one of several algorithms (such as the maximum

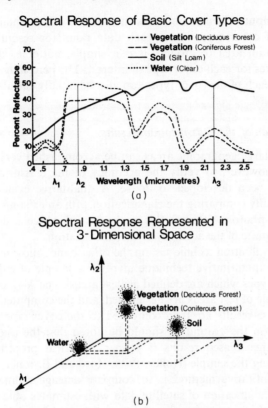

Figure 5.5 Schematic indicating the basis for computer classification of earth surface features

likelihood, parallelepiped, minimum distance to the means, ECHO, etc.) can be utilized for the classification itself. Different algorithms provide more or less accurate classifications and require varying amounts of computer time (Hoffer, 1979). Additional details are beyond the scope of this paper, but the entire subject is well documented in Swain and Davis (1978) and elsewhere.

5.4.3.4 Information Display and Tabulation

After the data are classified the results are stored on magnetic tape, and the analyst can display these results in a variety of map or tabular formats. Maps are usually in the form of 'thematic' maps in which each informational class of interest is displayed as a different colour or symbol.

Tabular outputs of classification results can be obtained easily, and are particularly useful for determining acreage. The analyst designates to the computer the *X–Y* coordinates representing the boundary of the area of interest. The computer then summarizes the number of data points in each of the categories of cover types. Since each data point (or resolution element) represents a certain area of ground, it is a simple matter to determine the number of hectares for each cover type of interest. The percentage of the entire area covered by each of the cover types can also be rapidly calculated.

5.4.3.5 *Evaluation of the Classification Results*

Classification of large geographic areas can be accomplished very rapidly. It is then desirable, however, to verify the accuracy of the classification. Several techniques have been developed to do so. A qualitative evaluation can be obtained by visually comparing the classification with an existing map of cover types with aerial photos. Although this method is subjective, it does provide a quick, rough estimate of the accuracy of the classification.

Quantitative evaluation techniques, on the other hand, allow more definitive evaluations. One quantitative technique involves a sample of individual areas of known cover types which are defined as 'test areas'. The $X-Y$ coordinates of a statistical sample of test areas are designated, and the computer tabulates the cover types. These results are then compared to the cover type known to be actually present on the ground. It should be noted that the use of test areas can produce rather biased results if proper statistical procedures are not followed in defining the sample (Hoffer, 1975; Hord and Brooner, 1976).

A second quantitative method is to compare acreage estimates obtained from computer classification of satellite data with estimates obtained by some conventional method, such as manual interpretation of aerial photos. If an adequate number of relatively large areas are summarized, a statistical correlation can be obtained.

5.5.4 Results of CAAT for Vegetation Mapping

Computer classification of LANDSAT and Skylab MSS data has shown that forestland, rangeland and agricultural lands can be distinguished from other cover types, and identified and mapped with a fairly high degree of accuracy, i.e., 80 to 95 per cent (NASA, 1972b, 1973a, 1973b, 1975, 1978; Heller, 1975; Hoffer, 1975; Hoffer and Staff, 1975; Dodge and Bryant, 1976; Williams and Haver, 1976; Hoffer and Fleming, 1978; Miller and Williams, 1978). LANDSAT classification estimates of total forest acreage were generally within ± 10 per cent of those obtained by forest survey of the US Forest Service (Dodge and Bryant, 1976; Aldrich, 1979). In one study, estimates of forest acreage for the state of Michigan obtained by classification of LANDSAT data were within two per cent of those obtained by the US Forest Service (Hoffer *et al.*, 1978). Roberts and Merritt (1977) obtained a forest acreage estimate for a nine-county area in Virginia that was within one per cent of the Forest Survey estimate. Thus, it would appear that quite accurate acreage estimates of forestland can be achieved by computer classification, at least over reasonably large areas. Such estimates were less accurate on smaller areas in each study.

In addition to being able to identify forested versus non-forested areas quite accurately, computer classification has the capability to differentiate between deciduous and coniferous cover types unless they occur in mixed stands, in which case the scanner system gives a spectral response that is approximately proportional to the mixture of cover types present but that also is influenced by variations in stand density (Dodge and Bryant, 1976; Williams and Haver, 1976; Hoffer and Fleming, 1978). Identification and mapping of individual forest species generally has been significantly less accurate, with results varying considerably (Hoffer and Staff, 1975; Hoffer, 1975; NASA, 1978). Spectral similarity among species often causes confusion, and variations in stand density as well as topographic effects cause significant differences in spectral response (Hoffer and Staff, 1975; Hoffer, 1975; Williams and Haver, 1976; Strahler *et al.*, 1978).

LANDSAT data can be obtained at regular intervals throughout the year and over a period of years on a worldwide basis, cloud cover permitting. This sequential coverage has made it possible to monitor the 'green wave' in the United States (Ashley and Rea, 1975; Blair and Baumgardner, 1977). Temporal differences in spectral response have also led to significant improvements in the classification of forest cover types (Williams, 1975; Kalensky and Scherk, 1975). Temporal changes in spectral response have been shown to be particularly important in identifying agricultural species (Steiner, 1970). Many studies have reported quite high (80 to 95 per cent) accuracy in classifying individual agricultural species (NASA, 1972b, 1973a, 1973b, 1975, 1978; Bauer, 1975; MacDonald and Hall, 1978). It should also be noted, however, that in many of these studies there were relatively few cover types present and field sizes were large, thereby providing a spectrally simple situation. The LACIE (Large Area Crop Inventory Experiment) project showed that temporal differences in spectral response are particularly important in accurately identifying wheat (MacDonald and Hall, 1978).

The availability of LANDSAT MSS data also raises the possibility of monitoring changes in the areal extent of vegetative cover. For example, Klankamsorn (1976) reported that a comparison of LANDSAT data obtained over Thailand in 1973 with aerial photos obtained in 1961 showed that the total area of forestland in that country had decreased from 55 per cent to 41 per cent. Miller and Williams (1978) have used LANDSAT data to evaluate land-use changes in Thailand, Taiwan, Nigeria and the Dominican Republic, particularly the conversion of forestland to agricultural use. A variety of computer analysis techniques to identify areas of land-use change have been developed and assessed, but the results of such 'change-detection analysis techniques' have been somewhat mixed (Weismiller *et al.*, 1977).

5.4.5 Summary of the Value of MSS Data for Vegetation Mapping and Assessment

The availability of LANDSAT data for most areas of the world, often for several dates encompassing different seasons and different years, offers great potential for mapping and monitoring vegetation. The scale of the imagery produced is small, and the spatial resolution of the LANDSAT scanner (0.46 ha) is such that this type of data should not be considered as a potential substitute for photographic data but rather as suitable for the first stage in a multistage analysis. Both manual and computer-aided analysis techniques have been shown to be useful. The decision as to which one is more appropriate depends largely on the problem to be solved and the degree of detail required. Use of data from different years offers significant potential for monitoring deforestation in many critical areas of the world.

5.5   RADAR SYSTEMS FOR VEGETATION MAPPING

5.5.1   Radar System Characteristics

Radar is the third major type of remote sensing system utilized for vegetation mapping. Since radar systems operate in the microwave portion of the spectrum, they offer certain advantages over both photographic and multispectral scanner systems.

The term 'RADAR' is an acronym for 'RAdio Detection And Ranging', which succinctly summarizes the characteristics of these systems. A radio signal is transmitted from the radar antenna, and the length of time required for the signal to travel to the ground and be reflected back to the antenna allows the distance of the object from which the signal was reflected to be determined. The signals are pulses of very short duration, the duration determining the across-track (range) resolution of the system (see Figure 5.6). Jensen *et al.* (1977) point out that a pulse lasting only 10^{-7} seconds is used in one system and produces a range resolution of 15 m. The along-track resolution is proportional to the width of the beam of the microwave signal, which is inversely proportional to the length of the antenna. The antenna length can be artificially enlarged and the resolution significantly improved through the use of synthetic aperture systems. A unique feature of synthetic aperture systems is that along-track resolution remains the same at all ranges, thereby making it possible to obtain high resolution imagery for objects many miles away. This is due to the fact that as the distance to the object increases the object remains in the antenna beam for a longer period of time. Thus, the effective length of the synthetic antenna is directly proportional to the range to the feature, and since the resolution is proportional to the length of the antenna but inversely proportional to the range, these two effects compensate for each other on synthetic aperture radar (Jensen *et al.*, 1977).

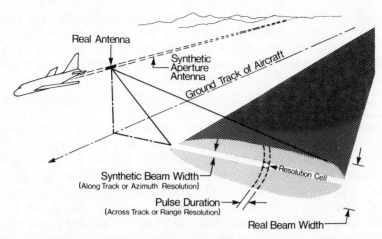

Figure 5.6 Schematic of the basic data collection characteristics of Side-Looking Airborne Radar systems (after Goodyear Aerospace Corp. Staff, 1971)

5.5.2 Interpretation of Radar Imagery

Although the all-weather and day-or-night capabilities afforded by radar systems are probably their most frequently cited advantages, they have several other unique capabilities for mapping and monitoring vegetation. To appreciate these advantages, one must understand some of the characteristics of radar systems and the data obtained from such systems. Key elements that must be considered in the interpretation of radar imagery are related to the radar system characteristics, which include wavelength, spatial resolution, look angle and polarization; the topographic characteristics of the area imaged, including the variations due to slope and aspect, and the effect of radar shadows; and the characteristics of the surface materials, including their geometric properties and dielectric properties (Hoffer, 1976b).

One of the most obvious characteristics of radar systems used in earth resource surveys is that they are side-looking (SLAR = Side Looking Airborne Radar) systems, viewing the terrain from an oblique angle. This enhances many characteristics of the terrain, a fact that has advantages and disadvantages. Areas behind tall terrain features (e.g., mountains) often are in a 'radar shadow' where there is no return of the radar signal, and the area is totally black on the imagery. Mountainsides and slopes facing the radar antenna provide a much higher return than areas of similar cover types on flat terrain. Areas with the same aspect and vegetative cover but with different slopes create different tones on radar imagery, thereby making it difficult to interpret the cause of such variations (Mathews, 1975; Hoffer, 1976b).

The side-look angle offers some advantages for vegetation mapping in that different physiognomic classes of vegetation can be enhanced on SLAR imagery. In forested areas, trees with large crowns cause a rough texture on the radar imagery, as do stands having a very low canopy closure. Brushland areas produce a finer texture. Thus, differentiation and mapping of major vegetative cover types can be achieved. Some agricultural crops can be accurately identified and mapped using SLAR imagery at selected times during the growing season (Morain and Simonett, 1966; Mathews, 1975; Bajzak, 1976; Ulaby and Burns, 1977).

In addition to differences in canopy texture, differences in the moisture content of vegetation significantly influence the radar signal, particularly in the shorter (i.e. K and X) wavelength bands. This is because the radar signal at these frequencies is strongly reflected by material with a high dielectric constant, and the dielectric constant is closely correlated with the moisture content. Therefore, vegetation having a high moisture content, or a fairly complete canopy of turgid vegetation, will produce a relatively light tone on the radar imagery (Morain and Simonett, 1966; Mathews, 1975; Rouse, 1977). The ability of radar to discriminate between agricultural fields with different levels of moisture is one of the primary reasons that certain crop species can be identified at specific times during the growing season. It also offers some hope for detecting stress conditions (Mathews, 1975).

One of the characteristics of radar which has often been misrepresented is its ability to 'see through' vegetative canopies. Longer wavelength systems, such as L-band, do have a significant potential for penetrating surficial cover, but the shorter wavelength systems, such as K-band, do not (Mathews, 1975). Since the spatial resolution of L-band radar is much poorer than that of the shorter wavelength K- or X-band systems, and since it has been primarily the K- or X-band systems that have been available to scientists for vegetation mapping, most of the data obtained thus far has come from K- or X-band systems. With these systems, in situations where the terrain does not cause significant variations, the geometric characteristics and moisture content of the vegetation itself are the dominant factors influencing the texture and tone of the radar image (Morain and Simonett, 1966, 1967; Barr and Miles, 1970; Mathews, 1975).

The polarization of the radar signal can also influence significantly the appearance of different vegetative types on radar imagery. In some cases the HH polarization gives better differentiation between vegetative cover types than the HV; in other cases the reverse is true (Morain and Simonett, 1966; Westinghouse, 1971; Lewis, 1973).

Another characteristic of radar data is the small scale involved and the resultant capability to cover large areas rapidly. One system, for example, is flown at a height of 20 000 ft, obtaining data at a scale of 1:250 000 and covering a swath about 12 miles in width (Francis, 1976). The unique capabilities of synthetic aperture radar systems are of particular interest, since relatively high spatial resolution can be achieved from great distances—e.g.,

10 m at a distance of 100 km (Jensen *et al.*, 1977). Very complex navigation and radar recording electronics are required in such systems in order to permit the extremely long antenna to be synthesized.

Radar systems operating in the K- and X-bands (approximately 0.83 and 3.0 cm wavelengths, respectively) have been the most commonly used for vegetation mapping. Such systems have been particularly useful in tropical regions because of the unique capability of radar to penetrate cloud cover. In 1968 the Darien Province of Panama was successfully mapped through the use of a K-band radar system, which obtained imagery of good quality in spite of the almost perpetual cloud cover over the area (Viksne *et al.*, 1969). Synthetic aperture radar systems were released from military classification in 1970, and since that time significant portions of the earth's surface have been mapped using an X-band system (Jensen *et al.*, 1977; Rouse, 1977). One of the most notable achievements of this system involved Project RADAM (van Roessel and deGodoy, 1974), in which SLAR data were obtained for the entire Brazilian Amazon Basin, an area of approximately 4×10^6 km^2, in less than a year's time. Land-use, vegetation, hydrologic, geologic and other types of maps were prepared from this imagery. In 1976 a radar survey of the entire country of Brazil was completed. Large portions of many other countries in Central and South America, Southeast Asia, Africa and the Far East have also been surveyed (Mathews, 1975; Francis, 1976; Jensen *et al.*, 1977).

5.5.3 Summary of Radar's Potential for Vegetation Mapping

Radar has the capability of obtaining small-scale data with relatively high resolution from high-flying aircraft or from spacecraft. Such data can be obtained reliably at specific stages in the growth and development of agricultural crops and forest vegetation types, regardless of cloud cover. Although topographic relief often causes unwanted variations in tone, many major vegetative cover types can still be differentiated and mapped. The physiognomic characteristics of vegetative cover types and the moisture content of the vegetation are the primary causes of differences in texture and tone on radar imagery. Changes such as recent clear-cutting in tropical forest regions are distinct on radar imagery (Mathews, 1975).

5.6 SUMMARY

It is clear that each of the three data collection systems discussed in this paper has unique advantages. LANDSAT provides a relatively economical way of obtaining sequential data at a scale and in a format that is appropriate for monitoring global vegetation, using computer-aided analysis. However, scanner systems operating in the optical wavelengths cannot obtain data in areas where there is persistent heavy cloud cover, whereas radar can. The capability of radar to provide data related to plant physiognomy offers a potential for differentiating among vegetative cover types and sizes that cannot be distinguished through the use of spectral data alone. The advantages of photographic data are that they provide a degree of detail that cannot be obtained by the other types of sensors. Thus, each type of sensor provides the capability of obtaining data that cannot be obtained in any other way. The type and degree of detail of the information needed must be carefully defined, after which the various sensor systems can be matched to the information required. Different analysis techniques must be utilized, depending on the sensor system involved, the scale of imagery obtained, and the degree of detail required. Both manual and computer-aided analysis techniques have distinct advantages and limitations that must be recognized in order to achieve maximum efficiency.

When one considers the various types of sensor systems and analysis techniques available, it is apparent that remote sensing technology offers a powerful and relatively economical tool for assessing the extent, characteristics and condition of the vegetation resources of the world.

5.7 REFERENCES

Adrien, P. M., and Baumgardner, M. F. (1977) LANDSAT, computers, and development projects. *Science*, **198**, 466–470.

Aldrich, R. C. (1971) Space photos for land use and forestry. *Photogram. Eng.*, **37**(4), 389–401.

Aldrich, R. C. (1975) Detecting disturbances in a forest environment. *Photogram. Eng. and Rem. Sens.*, **41**(1), 39–48.

Aldrich, R. C. (1979) Remote sensing of wildland resources; a state-of-the-art review, 1978. *U.S. Forest Service Report*, Fort Collins, Colorado. 131 pp.

Ashley, M. D., and Rea, J. (1975) Seasonal vegetation differences from ERTS imagery. *Photogram. Eng. Rem. Sens.*, **41**(6), 713–719.

Avery, T. E. (1977) *Interpretation of Aerial Photographs*. Burgess Publishing Co., Minneapolis, Minnesota. 392 pp.

Bajzak, D. (1976) Interpretation of vegetation types on side looking airborne radar and on thermal infra-red imagery. In: Hildebrant, G. (ed), *Proceedings of the Remote Sensing in Forestry Symposium held during the XVI IUFRO World Congress, Oslo, Norway*, 87–99. University of Freiburg, Freiburg, Germany.

Barr, D. J., and Miles, R. D. (1970) SLAR imagery and site selection. *Photogram. Eng.*, **36**(11), 1155–1170.

Bartolucci, L. A., Robinson, B. F., and Silva, L. F. (1977) Field measurements of the spectral response of natural waters. *Photogram. Eng. Rem. Sens.*, **43**(5), 595–598.

Bauer, M. E. (1975) The role of remote sensing in determining the distribution and yield of crops. *Adv. Agron.*, **27**, 271–304.

Bauer, M. E., Swain, P. H., Mroczynski, R. P., Anuta, P. E., and MacDonald, R. B. (1971) Detection of southern corn leaf blight by remote sensing techniques. *Proceedings of the 7th International Symposium on Remote Sensing of Environment*. University of Michigan, Ann Arbor, Michigan.

Blair, B. O., and Baumgardner, M. L. (1977) Detection of the green and brown wave in hardwood canopy covers using multidate, multispectral data from LANDSAT-1, *Agron. J.*, **69**, 808–811.

Colwell, R. N. (1956) Determining the prevalence of certain cereal crop diseases by means of aerial photography. *Hilgardia*, **26**(5), 223–286.

Colwell, R. N. (ed) (1960) *Manual of Photo Interpretation*. American Society of Photogrammetry, Falls Church, Virginia. 868 pp.

Colwell, R. N. (1968) Remote sensing of natural resources. *Sci. Am.*, **218**(1), 54–69.

Dodge, A. G. Jr., and Bryant, E. S. (1976) Forest type mapping with satellite data. *J. For.*, **74**(8), 526–531.

Draeger, W. C., Pettinger, L. R., and Benson, A. S. (1971) The use of small-scale aerial photography in a regional agricultural survey. *Proceedings of the 7th International Symposium on Remote Sensing of the Environment*. University of Michigan, Ann Arbor, Michigan.

Driscoll, R. S., and Coleman, M. D. (1974) Color for shrubs. *Photogram. Eng.*, **40**(4), 451–459.

Francis, D. A. (1976) Possibilities and problems of radar image interpretation for vegetation and forest types with particular reference to the humid tropics. In: Hildebrandt, G. (ed), *Proceedings of the Symposium on Remote Sensing in Forestry held during the XVI IUFRO World Congress, Oslo, Norway*, 79–86. University of Freiburg, Freiburg, Germany.

Fritz, N. L. (1967) Optimum methods for using infra-red-sensitive color films. *Photogram. Eng.*, **33**(10), 1128–1138.

Gates, D. M., Keegan, H. J., Schleter, J. C., and Weidner, V. R. (1965) Spectral properties of plants. *Appl. Optics*, **4**(1), 11–20.

Goodyear Aerospace Corporation Staff (1971) *Terrain imaging radar: Criteria for system selection*. Goodyear Aerospace Corp., Litchfield Park, Arizona. 64 pp.

Hasel, P. (1972) Michigan experimental scanner system. *Proceedings of the 4th Annual Earth Resources Program Review*, 34-1–34-13. NASA Johnson Space Center, Houston, Texas.

Hay, C. M. (1974) Agricultural inventory techniques with orbital and high-altitude imagery. *Photogram. Eng.*, **40**(11), 1283–1293.

Heller, R. C. (1970) Imaging with photographic sensors. Remote Sensing, with Special Reference to Agriculture and Forestry. National Academy of Science, Washington, DC. 424 pp.

Heller, R. C. (1971) Color and false color photography: its growing use in forestry. *Application of Remote Sensors in Forestry*, 21–36, IUFRO, Freiburg, Germany.

Heller, R. C., Technical Coordinator (1975) Evaluation of ERTS-1 data for forest and range-land survey. *United States Department of Agriculture Forest Service Research Paper PSW-112*, Pacific S.W. Forest and Range Experimental Station, Berkeley, California. 67 pp.

Heller, R. C. (1978) Case applications of remote sensing for vegetation damage assessment. *Photogram. Eng. Rem. Sens.*, **44**(9), 1159–1166.

Heller, R. C., Doverspike, G. E., and Aldrich, R. C. (1966) Identification of tree species on large scale panchromatic and color aerial photographs. *USDA Handbook 261*. 17 pp.

Hoffer, R. M. (1967) Interpretation of remote multispectral imagery of agricultural crops. LARS Vol. 1, *Research Bulletin 831*. Agricultural Experiment Station, Purdue, University, W. Lafayette, Indiana. 36 pp.

Hoffer, R. M. (1975) Computer-aided analysis of Skylab scanner data for land use mapping, forestry and water resource applications. *Proceedings of the Eleventh International Symposium on Space Technology and Science*, 935–941. Tokyo, Japan.

Hoffer, R. M. (1976a) Techniques and applications for computer-aided analysis of multispectral scanner data. *Proceedings of XVI IUFRO World Congress, Division VI*, 244–254. Oslo, Norway.

Hoffer, R. M. (1976b) Interpretation of radar imagery. *Fundamentals of Remote Sensing Minicourse Series*, Office of Continuing Education, Purdue University, West Lafayette, Indiana. 11 pp.

Hoffer, R. M. (1978) Biological and physical considerations in applying computer-aided analysis techniques to remote sensor data. In: Swain, P. H. and Davis, S. M. (ed), *Remote Sensing: The Quantitative Approach*, Chapter 5, 227–289. McGraw-Hill, NY, NY.

Hoffer, R. M. (1979) Computer-aided analysis of remote sensor data—magic, mystery or myth? *Proceedings of Remote Sensing for Natural Resources: An International View of Problems, Promises, and Accomplishments*, 156–179. University of Idaho, Moscow, Idaho.

Hoffer, R. M., and Fleming, M. D. (1978) Mapping vegetative cover by computer-aided analysis of satellite data. *Proceedings of the Workshop on Integrated Inventories of Renewable Natural Resources, General Technical Report RM-55*, Rocky Mountain Forest and Range Experiment Station, US Forest Service, Fort Collins, Colorado. 227–237.

Hoffer, R. M., and Johannsen, C. J. (1969) Ecological Potentials in Spectral Signature Analysis. In: Johnson, P. (ed), *Remote Sensing in Ecology*, Chapter 1, 1–16. University of Georgia Press, Athens, Georgia.

Hoffer, R. M., Noyer, S. C., and Mroczynski, R. P. (1978) A comparison of LANDSAT and forest survey estimates of forest cover. *Proceedings of the Fall Technical Meeting of the American Society of Photogrammetry*, 221–231. Falls Church, Virginia.

Hoffer, R. M. and Staff (1975) Natural resource mapping in mountainous terrain by computer analysis of ERTS-1 satellite data. *Agricultural Experiment Station Research Bulletin 919*. Purdue University, West Lafayette, Indiana. 124 pp.

Holter, M. R., Courtney, H. W., and Limperis, T. (1970) Research needs: the influence of discrimination data processing and system design. *Remote Sensing and Special Emphasis on Agriculture and Forestry*, 354–421. National Academy of Sciences, Washington, DC.

Hooley, R., Hoffer, R., and Morain, S. (1977) Estimating agricultural production by the use of satellite data: An experiment with Laotian data. *Am. J. Agric. Econ.*, **59**(4), 722–727.

Hord, R. M., and Brooner, W. (1976) Land use map accuracy criteria. *Photogram. Eng. Rem. Sens.*, **42**, 671–677.

Howard, J. A. (1971) Reflective foliaceous properties of tree species. In: *Applications of Remote Sensors in Forestry*, 127–146. IUFRO, Freiburg, Germany.

Hsu, K. S. (1978) The evaluation of LANDSAT data and analysis techniques for mapping tropical forest areas. PhD dissertation, Purdue University, West Lafayette, Indiana. 176 pp.

Jensen, H., Graham, L. C., Porcello, L. J., and Leith, E. N. (1977) Side-looking airborne radar. *Sci. Am.*, **237**(4), 84–95.

Kalensky, Z., and Scherk, L. R. (1975) Accuracy of forest mapping from LANDSAT computer compatible tapes. *Proceedings of the Tenth International Symposium on Remote Sensing of Environment*, 1159–1167. Ann Arbor, Michigan.

Kalensky, Z., and Wilson, D. A. (1975) Spectral signatures of forest trees. *Proceedings of the Third Canadian Symposium on Remote Sensing*, 155–171. Edmonton, Alberta, Canada.

Klankamsorn, B. (1976) LANDSAT-1 imagery application in forestry. In: Hildebrandt,

G. (ed), *Proceedings of the Remote Sensing in Forestry Symposium held during the XVI IUFRO World Congress, Oslo, Norway*, 227–233. University of Freiburg, Freiburg, Germany.

Kodak (1976) Kodak Aerochrome Infrared Film 2443. *Kodak Publication No. M-69*, Eastman Kodak Company, Rochester, New York. 11 pp.

Kourtz, P. H., and Scott, A. J. (1978) An improved image enhancement technique and its application to forest fire management. *Proceedings of the Fifth Canadian Symposium for Remote Sensing, Victoria, British Columbia.* 9 pp.

Landgrebe, D. A., and Staff (1967) Automatic identification and classification of wheat by remote sensing. *Research Progress Report 279*, Agricultural Experiment Station, Purdue University, West Lafayette, Indiana. 7 pp.

Langley, P. G., Aldrich, R. C., and Heller, R. D. (1969) Multistage sampling of forest resources by using space photography—an Apollo 9 case study. *Proceedings of the 2nd Annual Earth Resources Aircraft Program Review*, 19-1–19-21, Volume 2: Agr. Forest, and Sensor Studies. NASA MSC, Houston, Texas.

Lauer, D. T. (1971) Multiband photography for forestry purposes. In: *Application of Remote Sensors in Forestry*, 21–36, IUFRO, Freiburg, Germany.

Lauer, D. T., and Benson, A. S. (1973) Classification of forest lands with ultra-high altitude, small scale, false-color infra-red photography. *Proceedings of the IUFRO Symposium on Remote Sensing*, 143–162. Freiburg, Germany.

Lee, Y. J. (1975) Are clear-cut areas estimated from LANDSAT imagery reliable? *Proceedings of the NASA Earth Resources Survey Symposium*, 105–114. Houston, Texas.

Lewis, A. J. (1973) Evaluation of multiple-polarized radar imagery for the detection of selected cultural features. In: Holtz, R. (ed), *The Surveillant Science*, 297–313. Houghton Mifflin Co., Boston, Massachusetts.

Lowe, D. S., Polcyn, F. C., and Shay, R. (1964) Multispectral data collection program. *Proceedings of the Third International Symposium on Remote Sensing of the Environment*, 667–680. University of Michigan, Ann Arbor, Michigan.

MacDonald, R. B., and Hall, F. G. (1978) LACIE: An experiment in global crop forecasting. *Proceedings of the LACIE Symposium, Report No. OSC-14551*, 17–48. National Aeronautics and Space Administration, Johnson Spacecraft Center, Houston, Texas.

Manzer, F. E., and Cooper, G. R. (1967) Aerial photographic methods of potato disease detection. *Agricultural Experiment Station Bulletin 646*, University of Maine, Orono, Maine.

Marshall, J., and Meyer, M. (1977) A field evaluation of small-scale forest resource aerial photography. *IAFHE RSL Research Report 77-2*, University of Minnesota, St. Paul, Minnesota. 16 pp.

Mathews, R. E. (ed) (1975) *Active Microwave Workshop Report.* NASA SP-376, National Aeronautics and Space Administration, Washington, DC. 502 pp.

Miller, L. D., and Williams, D. L. (1978) Monitoring forest canopy alteration around the world with digital analysis of LANDSAT imagery. From *Proceedings ISP-IUFRO International Symposium on Remote Sensing Observation and Inventory of Earth Resources and Endangered Environment*, Freiburg, Germany. 41 pp.

Morain, S. A., and Simonett, D. S. (1966) Vegetation analysis with radar imagery. *CRES Report 61–9*, University of Kansas, Lawrence, Kansas. 18 pp.

Morain, S. A., and Simonett, D. S. (1967) K-Band radar in vegetation mapping. *Photogram. Eng.*, **33**(7), 730–740.

Murtha, P. A. (1972) A guide to air photo interpretation of forest damage in Canada. *Canadian Forestry Service Publication No. 1292*, Ottawa, Canada. 63 pp.

158 *The role of terrestrial vegetation in the global carbon cycle*

Murtha, P. A. (1978) Remote sensing and vegetation damage: a theory for detection and assessment. *Photogram. Eng. Rem. Sens.*, **44**(9), 1147–1158.

NASA (1972a) *Earth Resources Technology Satellite: Data Users Handbook.* National Aeronautics and Space Administration, Goddard Space Flight Center, Greenbelt, Maryland.

NASA (1972b) *Earth Resource Technology Satellite—I Symposium Proceedings. Publication X-650-73-10.* National Aeronautics and Space Administration, Goddard Space Flight Center, Greenbelt, Maryland.

NASA (1973a) Symposium on significant results obtained from the earth resources technology satellite—1. *Publication X-650-73-127.* National Aeronautics and Space Administration, Goddard Space Flight Center, Greenbelt, Maryland.

NASA (1973b) Third earth resources technology satellite symposium. *NASA SP-357.* National Aeronautics and Space Administration, Goddard Space Flight Center, Greenbelt, Maryland.

NASA (1975) *Proceedings of the NASA Earth Resources Survey Symposium.* National Aeronautics and Space Administration, Johnson Space Center, Houston, Texas.

NASA (1978) *Skylab EREP Investigations Summary. NASA SP-399.* National Aeronautics and Space Administration, Johnson Space Center, Houston Texas. 386 pp.

Reeves, R. G. (ed) (1975) *Manual of Remote Sensing.* American Society of Photogrammetry, Falls Church, Virginia. 2144 pp.

Rhode, W. G., Lo, J. K., and Pohl, R. A. (1978) EROS Data Center LANDSAT digital enhancement techniques and imagery availability, 1977. *Can. J. Rem. Sens.*, **4**(1), 63–76.

Roberts, E. H., and Merritt, N. E. (1977) Computer-aided inventory of forestland. *Type III Final Report, USDA Forest Service,* 65–74. Rocky Mountain Forest and Range Experimental Station, Fort Collins, Colorado.

Rouse, J. W. Jr. (ed) (1977) *Workshop Report.* Microwave Remote Sensing Symposium, National Aeronautics and Space Administration, Johnson Space Center, Houston, Texas. 20 pp.

Sayn-Wittgenstein, L., and Aldred, A. H. (1967) Tree volumes from large-scale photos. *Photogram. Engin.*, **33**(1), 69–73.

Sinclair, T. R., Hoffer, R. M., and Schreiber, M. M. (1971) Reflectance and internal structure of leaves from several crops during a growing season. *Agron. J.*, **63**, 864–868.

Smedes, H. W., Pierce, K. L., Tanguey, M. C., and Hoffer, R. M. (1970) Digital computer terrain mapping from multispectral data. *J. Spacecraft and Rockets*, **7**(9), 1025–1031.

Smith, J. T. (ed) (1968) *Manual of Color Aerial Photography.* American Society of Photogrammetry, Falls Church, Virginia. 550 pp.

Steiner, D. (1970) Time dimension for crop surveys from space. *Photogram. Eng.*, **36**(2), 187–194.

Strahler, A. H., Logan, T. L., and Bryant, N. A. (1978) Improving forest cover classification accuracy from LANDSAT by incorporating topographic information. *Proceedings of the 12th International Symposium on Remote Sensing of the Environment,* 927–942. University of Michigan, Ann Arbor, Michigan.

Swain, P. H., and Davis, S. M. (ed) (1978) *Remote Sensing: The Quantitative Approach.* McGraw-Hill, Inc. NY, NY. 396 pp.

Ulaby, F. T., and Burns, G. (1977) The potential use of radar for crop classification and yield estimation. *Proceedings of the Microwave Remote Sensing Symposium.* National Aeronautics and Space Administration, Johnson Space Center, Houston, Texas.

van Roessel, J. W., and deGodoy, R. C. (1974) SLAR mosaics for Project RADAM. *Photogram Eng.*, **40**(5), 583–595.

Viksne, A., Listen, T. C., and Sapp, C. D. (1969) SLR reconnnaissance of Panama. *Geophysics*, **34**, 54–64.

Weismiller, R. A., Kristof, S. J., Scholz, D. K., Anuta, P. E., and Morain, S. A. (1977) Change detection in coastal zone environments. *Photogram. Eng. Rem. Sens.*, **43**(12), 1533–1539.

Westinghouse Corp. Staff (1971) *Side Look Radar.* Westinghouse Corporation. 46 pp.

Williams, D. L. (1975) Computer analysis and mapping of gypsy moth defoliation levels in Pennsylvania using LANDSAT-1 digital data. *Proceedings of the NASA Earth Resources Survey Symposium.* 167–177. NASA TMX-58168, JSC-09930, Houston, Texas.

Williams, D. L., and Haver, G. F. (1976) *Forest Land Management by Satellite: LANDSAT-derived Information as Input to a Forest Inventory System.* National Aeronautics and Space Administration, Goddard Spaceflight Center, Greenbelt, Maryland. 36 pp.

Yost, E. F., and Wenderoth, S. (1967) Multispectral color aerial photography. *Photogram Eng.*, **33**(9), 1020–1033.

The Role of Terrestrial Vegetation in the Global Carbon Cycle:
Measurement by Remote Sensing
Edited by G. M. Woodwell
© 1984 SCOPE. Published by John Wiley & Sons Ltd

Chapter 6

Remote Sensing for Monitoring Vegetation: An Emphasis on Satellites

F. C. BILLINGSLEY

Jet Propulsion Laboratory, California Institute of Technology, Pasadena, California, USA

ABSTRACT

The use of satellite imagery, especially the imagery from LANDSAT, has been used widely in mapping forested areas of the earth. The details of LANDSAT operation are summarized and the limitations of the system discussed. The applications of satellite imagery are under development, will increase substantially in the future, and include the possibility of measuring changes in the vegetation of the earth.

6.1 INTRODUCTION

LANDSAT imagery has proved to be very effective in both mapping and measuring areas of forests (University of Michigan, 1979), and there is substantial reason to believe that it can be used in measurements of changes in the storage of carbon in forests globally. The characteristics of forests that are used for mapping include stand boundaries, height, density, species, extent of dead and dying timber and site quality (NASA, 1976). The possibility exists, of course, that improvements can be made in the design of future satellites. I have chosen to emphasize the criteria to be used in these developments in addition to the use of the existing systems.

Several important factors must be considered in designing remote sensing instruments for this purpose. Some of the most important are:

(a) Spectral bandwidth. The bandwidth is normally selected to be wide enough to obtain sufficient energy and narrow enough to detect the spectral features of interest.
(b) Spectral selection. If complete spectrum sampling is not employed, the bands will be placed to be coincident with spectral features of interest.
(c) Spectral resolution. This is determined by the spectral spacing between band centres, not by the spectral bandwidth.

(d) Temporal resolution. The time of an image or the time between two images. The question is whether the instrumentation is to be used for routine, repeated monitoring, or to provide information on episodic events. The answer will be a factor in determining the appropriate orbit of the satellite and the timing of the imagery.

(e) Intensity resolution. The utility of finer and finer intervals of measurement of intensity is ultimately limited by the noise of the sensor itself and by the inherent variability in the scene.

(f) Accuracy. The atmosphere often interferes with reception. In addition, there may be a need for absolute calibration of the instrument in addition to normalizing between spectral bands.

(g) Correlation with other data or maps. Remotely sensed information will seldom be all the information required. The data must be prepared and transmitted to the user in a form that is useful in comparisons with other data.

(h) Data delay. Thoughtful instrument design may allow the production of data in forms that minimize the time needed to process it on the ground.

A large part of the instrument designer's and scientist's joint task is mutually to determine a suitable world model which can be solved using available data, and the sensor system to obtain that data.

6.2   DATA SAMPLING

6.2.1   Spectral Sampling Considerations

Approximately 87 per cent of the solar energy that reaches the earth is received in the spectral range of 0.3 to 3.0 micrometres (μm) and most investigations of remote sensing have considered this range (NASA, 1972; Kondratyev *et al.*, 1973; NASA, 1975a; Tucker and Maxwell, 1976; Landgrebe *et al.*, 1977; Ecosystems International, 1977; CORSPERS, 1976; Tucker, 1978a, 1978b; Rao *et al.*, 1978; Abrams *et al.*, 1977). A summary is given in Table 6.1. Figure 6.1 shows some of the bands recommended for investigation, the bands included in various sensors and solar irradiance as seen at sea level. The bands which are normally considered useful are those of low atmospheric attenuation. This requirement eliminates the regions of 0.75 to 0.8, 0.95 to 1.0, 1.1 to 1.18, 1.3 to 1.55, 1.8 to 2.0 and 2.3 to 3 μm. The range of 0.3 to 0.4 μm is usually not considered available because of large atmospheric effects, but it may be a good atmospheric sensing band.

Within these ranges, the chlorophyll absorption (0.47 μm) is sensed only by the Coastal Zone Color Scanner (CZCS); the water absorption band (0.75 to 0.8 μm) is sensed by part of a thematic mapper band and may therefore be a confusion factor as atmospheric water varies, the 1.0 to 1.1 μm band is part of

Table 6.1　Spectral Happenings

Wavelength (μm)	
0.3 –0.35 ⎱ 0.35–0.42 ⎰	In combination for atmospheric scattering and haze
0.43–0.45	Chlorophyll absorption (CZCS)
0.40–0.5	Carotenoid absorption
0.45–0.52 [a]	Bathymetry in less turbid waters; soil/vegetation differences; deciduous/confierous differentiation; soil type discrimination
0.51–0.53	Chlorophyll reflection (CZCS)
0.52–0.60 [a]	Indicator of growth rate and vegetation vigour because of sensitivity to green reflectance peak at 0.55 μm; sediment concentration estimation; bathymetry in turbid waters
0.54–0.56	Gelbe stuffe in water (CZCS)
0.63–0.69 [a]	Chlorophyll absorption/species differentiation; one of best bands for crop classification. Vegetation cover and density; with the 0.53–0.60 μm band it can be used for ferric iron detection; ice and snow mapping
0.6563	Chlorophyll luminescence line
0.66–0.68	Chlorophyll absorption (CZCS)
0.7–0.8	Chlorophyll reflectance rise; ground cover variations (CZCS)
0.76–0.9 [a]	Water body delineation; sensitive to biomass and stress variations. General observations with low atmospheric effects
0.82–0.9	Limonite absorption
1.4–1.5	Leaf turgidity
1.55–1.75 [a]	Vegetation moisture conditions and stress; snow/cloud differentiation; may aid in defining intrusives of different iron mineral content
1.9–2.0	Leaf turgidity
2.08–2.35 [a]	Distinguish hydrothermally altered zones from non-altered zones/mineral exploration; soil type discrimination
2.05–2.15 ⎱ 2.15–2.25 ⎰	In combination, for altered rock discrimination
2.2, 3.5, 3.9, 4.8, 13.2	Narrow bands for grey body, surface temperature
4.5–5.5	With 10 μm, for vegetative temperature, atmospheric water absorption
9.5–10	Ozone absorption
10.4–12.5 [a]	Surface temperature measurement; urban versus non-urban land use separation; burned areas from water bodies (CZCS)

a Thematic mapper bands.

the Multispectral Scanner (MSS) 0.8 to 1.1 μm band, and the 1.2 to 1.3 μm band is not covered at all.

Because the Thematic Mapper 0.45–0.52 μm band includes a chlorophyll band, further sensitivity may be obtained by widening the sensor band to include also the 0.44 μm chlorophyll band. The Thematic Mapper 0.76–0.9 μm band should be split to separate the water band from the rest to eliminate the water confusion and provide an adjacent band for normalization.

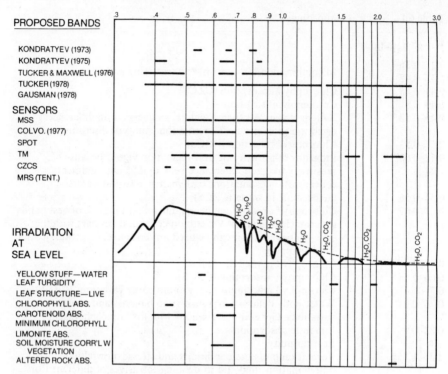

Figure 6.1   Spectral band placement in the 0.3–3 μm range

A multiband linear array sensor may be launched in the mid-1980s. It will be the first of a new generation of multispectral sensors—a 'pushbroom' sensor instead of a mechanical scanner (Thompson, 1979; Colvocoresses, 1979). For each band in a pushbroom sensor, a separate detector is provided for each picture element desired in the cross-orbit direction. This in turn provides a dwell time for sensing each element of the ground longer than that of a scanner by a factor of about 3000 for a 3000 pixel line, and provides an increase in signal/noise (S/N) ratio of approximately the square root of this amount, or about 50:1. Advantage may be taken of this higher S/N ratio by increasing the number of intensity levels or by sensing with narrower spectral bands (Collins, 1978; DelGrande, 1975). In particular, it should be possible to use the narrow chlorophyll absorption bands near 0.44 and 0.67 μm with an adequate S/N ratio and good spatial resolution. In addition, it may be possible to form images in the bands of high atmospheric attenuation or in the 0.3 to 0.35 and 0.35 to 0.43 μm bands. These latter images could be used in conjunction with other bands to estimate and correct for effects of atmospheric scattering.

Linear array sensors are limited now to a long wavelength cutoff of about 1 μm. When longer wavelength response can be provided, division of the

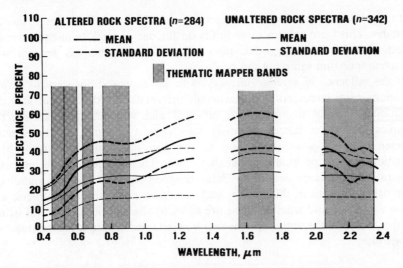

Figure 6.2 The position of the thematic mapper bands relative to average rock spectra and the plus or minus one standard deviation limits for hydrothermally altered rocks versus unaltered rocks. These figures were compiled from unpublished data furnished by Dr A. Kahle of the Jet Propulsion Laboratory in Pasadena, CA, Dr Larry Rowan of the US Geological Survey and other colleagues. (From Salomonson, 1978)

'2.2 μm' band into 2.05 to 2.18 and 2.18 to 2.35 μm sub-bands or even finer would allow further discrimination of, for example, altered from non-altered rocks (see Figure 6.2).

Additional bands of atmospheric transmittance occur between 3.4 to 4 μm and between 4.7 to 5 μm. These bands have not been given serious consideration because of the very low value of the reflected energy available and because of appreciable interference from thermally emitted energy from the earth. Within this range, narrow bands centred at 3.5, 3.9 and 4.8 μm avoid both atmospheric absorption regions and the wavelengths associated with anion groups in common minerals, and thus can serve as grey-body regions for the calculation of surface temperature (Del Grande, 1975). Conversely, other bands in this range may serve as indicators of these effects.

Temperature differences appearing in the thermal range have been associated with vegetation stress (Idso *et al.*, 1977; Millard *et al.*, 1978). In addition, it has been suggested by Vincent (1973) and others that the thermal band might be divided into several bands for some types of rock separations. Because of the presence of an ozone-absorbing band between 9.6 and 10 μm, it is suggested that this band be excluded.

A single pixel over vegetated ground usually records signals from several sources such as plants and soil (Tucker and Miller, 1977; Richardson and Wiegand, 1977a, 1977b, 1979; Westin and Lemme, 1978). Precise spectral features of vegetation may not be reproduced faithfully at pixel sizes at satellite altitudes. This factor places clear limits on the use of satellite imagery for the direct identification of ground materials until such time as sensors with complete spectrum sampling can be flown.

If the influence of satellite imagery is to be improved, answers to several questions seem important. First, whether the needs would be better served if additional spectral features were surveyed and whether sufficient spectral samples should be taken to satisfy the Nyquist criterion in the spectral domain. Second, whether sensing in the bands sensitive to atmospheric variation would provide normalizing data. Third, whether the potential spectral modifications caused by data compression would be less significant than other vagaries in the system, such as sensor noise and atmospheric and scene variations and whether these are offset by the additional spectral bands. Fourth, whether the entire scheme can be contained within a reasonable data bandwidth.

6.2.1.1   Spectral Data Analysis

Spectral analysis generally takes the form of multispectral classification, in which classification is done by comparing the sample measurement vector to the set of vectors representing all possible classes of known materials and, by using one of several methods, determining which of the knowns it most nearly matches. This subject, treated extensively in the literature, will not be pursued here. Rather, I will consider some of the effects of noise from a generic point of view.

Consider first the probability of correct identification of a pixel from the interior of a field. Reflectivity from such a field will vary with noise in the sensor (NEDR) and inherent variation in the field itself. The combined effect will produce a finite probability of misclassification. The accuracy of the procedure is greatest of course when the classes are well defined; the accuracy drops as noise increases. The drop in accuracy of classification with increasing noise has been defined by Ready *et al.* (1971) and is shown in Figure 6.3.

The effect may be estimated (Billingsley, 1981) (Figure 6.4). The probability of a sample being within the class limits can be derived by assuming that an ensemble of noise-free signals can have equal probability of being anywhere within the range and by adding a Gaussian noise with a distribution equal to σ. The probability distribution of the signal plus noise is found by convolving the probability distribution of the signal with that of the noise. The probability of correct class assignment is then found by integrating the probability distribution between appropriate limits representative of the class boundaries (Friedman, 1965). The resulting curve is given in Figure 6.5.

Sources of noise will be the scene itself and the sensor. A number of pixel measurements may be averaged to reduce the noise before classification. This

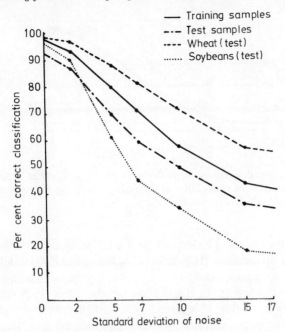

Figure 6.3 Effect of random noise on the accuracy of multispectral classification. (From Ready *et al.*, 1971)

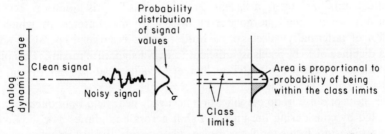

Figure 6.4 Effect of noise on the probability of correct multispectral classification

final noise figure may be compared to the width of the class to give β, from which the probability of correct classification may be estimated by using the Classification Error Estimator shown in Figure 6.6.

6.2.2 Spatial Sampling

The scene noise may itself be useful as a measure of texture. Texture is ordinarily described by terms such as: smooth, fine, rough and coarse. For digital processing, these terms are insufficient, and the spatial distribution of

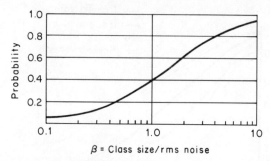

β = Class size/rms noise

Figure 6.5   Given a signal uniformly probable over the dynamic range. Gaussian noise of value $= \sigma$. The curves show probability of correctly recognizing a class corresponding to the noise free signal as a function of the ratio $\beta =$ class size/σ

the tones (brightness) of the pixels must be used. The specific attributes must be specified by the investigator (Haralick, 1975; Weszka and Rosenfeld, 1975).

The tradeoff is in sensor pixel size: (1) Large–low noise leading to better multispectral classification, poor delineation of edges, low amplitude texture information, larger fields required; (2) small–poorer multispectral classification, better edge definition, higher amplitude texture, smaller fields allowed, appreciably higher data production rate.

A further topic is 'how far away from a brightness discontinuity (e.g., edges of different ground cover) must samples be taken to avoid contamination (signature mixture) from adjacent ground areas?' This question becomes important where small homogeneous areas are encountered in which the fraction of potentially misclassified border pixels is large. Experience shows that a distance of $1–1\frac{1}{2}$ pixels is sufficient to avoid signature mixing. This places clear limits on the use of satellite imagery available at present in mapping areas as small as 1 ha or less.

The effect of noise upon classification accuracy near field boundaries may be estimated by considering the intensity shift across boundaries as compared to the classification limits and the noise components (including sensor noise) within a 'uniform' field. A trace cross-section between two fields is shown in Figure 6.7. As before, finite noise components will cause a finite probability of misclassification even if the average value is well within the classification limits, S. As the border of a field is approached the average value will change toward that of the adjacent field (Figure 6.8).

A final observation is that a shift of the signature in any one band will occur if that band is misregistered to the others, so that the pixel value shift is a combination of pixel placement and border displacement due to misregistration. The use of multitemporal data for classification implies that the spectral bands have equal importance and validity, since weighting factors are not generally applied on a band-to-band basis in accordance with their

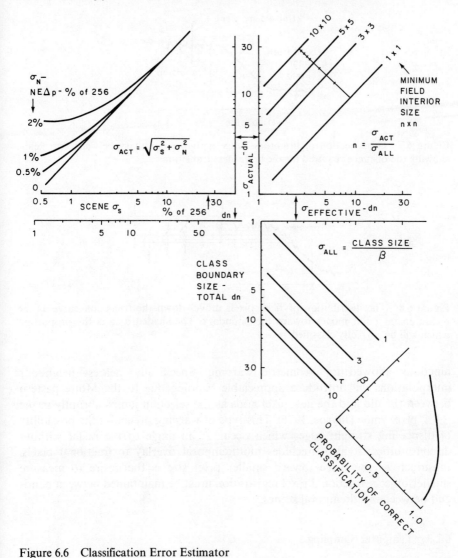

Figure 6.6   Classification Error Estimator

supposed utility. Therefore, all spectral bands, independent of date, may be considered to be equally valid, and the total misregistration effect may be considered to be the root-sum-square combination of the misregistrations of all of the bands. The effects have been modelled for an average field aspect ratio (r) and an average noise $(3 < \beta < 5)$. The loss in accuracy ΔP for a displacement of d pixels is given in Figure 6.9.

This analysis also has implications for the type of interpolation

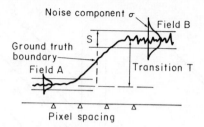

Figure 6.7   Cross-section of brightness trace across a boundary between two fields, showing the distance required for the brightness transition

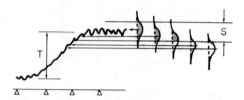

Figure 6.8   The distribution of 'field' pixels moves down the transition curve as the measurement point moves toward the boundary. The shaded area is the proportion which will be correctly classified

functions allowed for geometric warping. Specifically, nearest neighbour interpolations will introduce appreciable banding due to the Moire pattern between the old and the new pixel grids as the selection jumps abruptly at the $\pm 1/2$ pixel value (Jayroe, 1976). This type of warping precludes the possibility of displaying straight edges which occur at an angle to the raster without discontinuities; it also precludes multitemporal overlay to fractional pixels. Inasmuch as the drive toward smaller pixel size is the desire to measure smaller objects, fractional pixel registration must be maintained between bands and between multitemporal scenes.

6.2.3   Temporal Sampling

In sampling of vegetation over time the measurements sought are often the increase in the canopy cover early in the growing season, flowering or ripening, any stress conditions, and the abrupt change at harvest (Kauth and Thomas, 1976; Rabchevsky, 1977; Richardson and Wiegand, 1979; Wiegand *et al.*, 1979). Measurements at selected stages cannot be planned because a given species may be in different growth stages within the same region and cloud cover may obscure some observations. Repeated sensing within short intervals will be required to identify changes in signatures.

Repeat coverage will be obtained when the satellite completes a cycle and

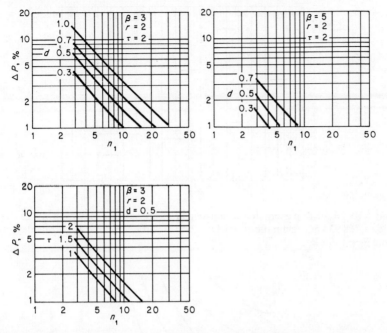

Figure 6.9 Effect of misregistration on multispectral classification accuracy—
Loss of classification accuracy due to misregistration of one band, for various parameter combinations.

β = class size/σ of noise.

r = Field shape ratio, long/short sides.

τ = 10–90 per cent transient distance.

n_1 = length of short side, pixels.

d = displacement, pixels.

ΔP = loss in probability

again traces a given orbit position, and when the swaths of nominally different orbits overlap. Within the altitude ranges normally considered (700–1000 km), there are relatively few orbits which are simultaneously sun synchronous and either progressing (such as LANDSAT 1, 2, 3) or skip (with adjacent orbits occurring approximately midway in the cycle) (LANDSAT 4). One such orbit pattern is shown in Figure 6.10. This has overlap coverage to the west and east of 8 and 11 days respectively, for a total cucle of 19 days.

The advantages of a skip orbit will only be realized if the swath is wide enough to give appreciably overlap with the adjacent swath down to the latitudes of interest. It can be shown that:

$$\frac{S_0}{W} = \frac{1 - C/2}{\cos \phi \cos H}$$

where S_0 equals trace spacing at the equator, W equals swath width

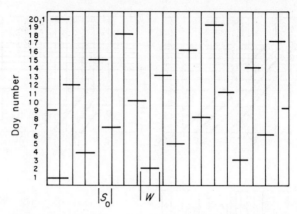

Figure 6.10 Repeat pattern for a sun-synchronous satellite having a 19-day cycle, altitude—768.93 km, inclination—98.47°, 273/19 orbits/cycle, $14\frac{7}{9}$ orbits/day, trace spacing $S_0 = 146.8$ km

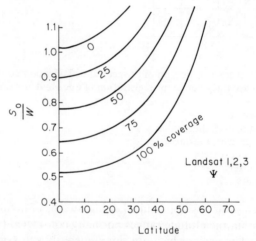

Figure 6.11 Minimum latitude for a given per cent coverage as a function of S_0/W, for LANDSAT 1, 2, 3, orbit parameters

perpendicular to the trace, C equals fractional coverage (1.0 for 50 per cent sidelap, 0 for just touching swaths), ϕ equals spacecraft latitude at any point and H equals the heading of the spacecraft relative to the rotating earth. A plot of this equation is given in Figure 6.11.

The additional data are not obtained without penalty. A decrease in S_0 will require more orbits to complete the cycle. An increase in the swath width requires a greater scan angle for a given altitude and produces greater distortion at the edges of the images. If the distortion cannot be removed by ground processing, it must be minimized during the sensing by increasing the altitude. The extra swath width also requires a higher data acquisition and handling rate.

Immediate attention may be given to episodal events only if the spacecraft is there at the time or if the sensor can be pointed at the event from another spacecraft position. This conjunction may also be provided by geosynchronous satellites, but at the cost of launching problems and attainable sensor resolution. Delay before the next acquisition in a low earth orbit may be minimized through the use of a skip orbit similar to the one described above and the use of a pointable sensor.

6.2.4 Accuracy of Data

The radiance measured by the spacecraft is a combination of the energy reflected from the ground (attenuated by the atmosphere) and the radiance of the atmospheric path. The path radiance, in turn, is caused by the illumination of atmospheric components by sunlight, by other components of the atmosphere and by light reflected to the ground. Since the distribution of illuminable particles is unknown and is variable both vertically and horizontally, and since ground reflectivity is spatially quite variable, path radiance has not been modelled adequately. Due to the combined effects of attenuation and path radiance, a given pixel as seen at the spacecraft may appear either lighter or darker than it would at the ground. Sensors are not yet available to measure atmospheric effects.

These effects have typically been estimated by one of two methods, each of which assumes the effects to be spatially invariant: (1) assuming that the darkest pixel in an image has some known low reflectivity, subtract this value from all pixels; or (2) assume that a series of brightness measures made just inside and just outside of a cloud shadow will bear a linear relationship to each other (Calspan, 1976). This relationship may be used to calculate attenuation and brightness. Atmospheric effects modelled by Mie or Rayleigh scattering or a combination thereof will be relatively slow varying functions of wavelength. In this case a ratio image will tend to remove atmospheric effects. although even here dark level subtraction will improve the ratio results. A declaration by the various disciplines as to the necessity of absolute calibration of spacecraft instruments and the need to measure atmospheric effects would help to guide instrument designers.

6.2.5 Correlation of Data

Miller (1979) concluded that it is time to develop techniques for measuring forest depletion by means of LANDSAT. By the time a programme to do so can be set in motion we will have over ten years of LANDSAT data as a

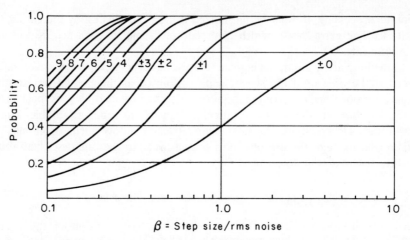

β = Step size/rms noise

Figure 6.12   Given a signal uniformly distributed over the quantization intervals. Given a Gaussian noise value = σ. The curves show probability of correctly assigning a digital value corresponding to the noise-free signal within ± 0, ± 1,... ± 9 DN (inclusive) as a function of the ratio β = step size/σ (from Billingsley, 1975)

base. In view of the possibilities of deterioration of the older data, it is time to plant its retrieval or at least obtain assurance of its retention. Producing a history of forest depletion will not be simple, however, because of the need to register vast quantities of data on a world-wide scale. The development of mosaicking and registration techniques by NASA could be accelerated and made more available for general use if such a need is acknowledged.

6.2.6   LANDSAT-D and the Thematic Mapper

LANDSAT-D (LANDSAT-4) was launched in 1982, as part of a complete, end-to-end, highly automated earth monitoring system, and is thus a major step forward in remote global sensing (CORSPERS, 1976; Salomonson, 1978).

The LANDSAT-4 spacecraft has improved pointing accuracy and stability characteristics, namely, 0.01 degree and 10^{-6} degrees per second, respectively. It was launched into a sun-synchronous orbit near 705 km to provide coverage every 16 days. The principal instruments on the spacecraft are the Thematic Mapper (TM) and the Multispectral Scanner (MSS). The MSS provides the same capability as the MSS on LANDSAT-3, but the Thematic Mapper has capabilities superior to those of the MSS. It has seven spectral bands: 0.45 to 0.52, 0.52 to 0.60, 0.63 to 0.69, 0.76 to 0.90, 1.55 to 1.75, 2.08 to 2.35 and 10.4 to 12.5 μm. The six visible and near infra-red bands provide a 30-m instantaneous field of view (IFOV), and the thermal band provides a 120-m IFOV. The TM is an eight-bit system; i.e., it has 256 levels. The data rate from the TM is

Table 6.2 Thematic mapper and multispectral scanner subsystem characteristics for LANDSAT-4 (from Salomonson, 1978)

	Thematic Mapper (TM)		Multispectral Scanner Subsystem (MSS)	
	Micrometers	Radiometric Sensitivity (NEDR)	Micrometers	Radiometric sensitivity (NEDR)
Spectral band 1	0.45–0.52	0.8%	0.5–0.6	0.57%
Spectral band 2	0.52–0.60	0.5%	0.6–0.7	0.57%
Spectral band 3	0.63–0.69	0.5%	0.7–0.8	0.65%
Spectral band 4	0.76–0.90	0.5%	0.8–1.1	0.70%
Spectral band 5	1.55–1.75	1.0%		
Spectral band 6	10.40–12.50	0.5 K(NE ΔT)		
Spectral band 7	2.08–2.35	2.4%		
Ground IFOV		30 M (except band b = 120 m)	82 M (bands 1–4)	
Data rate		83 mB/s	15 mB/s	
Quantization levels		256	64	
Weight		227 kg	81 kg	
Size	1.1 × 0.7 × 2.0 m		0.35 × 0.4 × 0.9 m	
Power	320 watts		42 watts	

approximately 85 megabits per second. Table 6.2 lists the pertinent characteristics of the instruments.

6.2.7 Microwave Sensing

Approximately 1200 LANDSAT scenes will be required to cover the tropical forest area of the earth. Many of these areas have never been imaged because of pervasive cloud cover. LANDSAT therefore cannot provide all the needed data even if the required frequency of observation is only once a year (CORSPERS, 1977).

An active microwave system with nearly all-weather capability could help fill this gap. The primary need is to distinguish between forested and cleared land, a classification task that appears to be within the capability of existing radar technology. Canopy penetration is not required, and a relatively short wavelength system with a 25-m resolution should be adequate. Polarization does not appear to be critical.

However, much more research must be done before definite statements can be made. A number of research topics have been suggested (Chafaris, 1978):

(1) Measurement of stand density. Differences in stand density may show up on radar that are more reliable than LANDSAT MSS data alone.

(2) Area delineation. Synthetic aperture radar (SAR) may be useful in differentiating forest lands from brush and pasture. Confusion of these features was a problem in the Great Lakes Watershed mapping project of the Environmental Protection Agency (EPA).
(3) Species differentiation. The SAR, in conjunction with the MSS, may assist in differentiating tree species which have spectrally similar signatures in the visible and near-visible infra-red ranges of the spectrum.
(4) Tropical forest inventory. The SAR might become a prime sensor in measuring the areal extent of cloud-covered tropical forests.
(5) Tree height. SAR shadowing may be related to tree height (assuming appropriate incidence angles) and thus aid in ascertaining timber volume.
(6) Detection of tree stress and tree susceptibility to fire. The detection of stress is not a straightforward problem. It is influenced by terrain slopes and the direction of slope as well as the morphology of the forest on the radar backscatter amplitude.

A synthetic aperture radar (SAR) was flown on SEASAT. It operated at 1.3 GHz and an off-nadir angle of 17–23°. The swath width was 100 km, and processing yielded 25×25 m pixels. Although the mission met an untimely end, considerable data were obtained over both water and land. Some of these data have been processed.

The registration of SAR imagery to maps and to LANDSAT images is not a simple task, due to the difference in the appearance of tie points and the relief distortion due to the large off-nadir angle. Efforts are under way to solve these difficulties.

The SEASAT SAR was designed primarily to provide data on ocean waves, coastal regions and sea ice. A SAR to provide data on land features, the Shuttle Imaging Radar (SIR-A), was flown on the second Shuttle flight test (OSTA-1) in November, 1981. This experiment used a modified SEASAT SAR to provide imagery in the L-band (23 cm or 1.3 GHz) region, with resolution compatible with LANDSAT imagery. SIR-A operated at an off-nadir angle of 50°, with a 50 km swath. Processing yields 40 m pixels from an optical signal film recorded on-board. SIR-B is expected to be launched in the summer of 1984. It will allow digital acquisition of data at selected incidence angles between 15 and 60°. This method represents a significant step in understanding optimum viewing angles for various applications.

The brightness of the radar image is proportional to the backscatter cross-section of the surface. This, in turn, is related to surface roughness, topography, dielectric constant and vegetation cover. Thus, radar data correlated with LANDSAT images will provide more information than either sensor alone. Because of the responsiveness of radar to so many variables, however, more research is required to determine the best combination of parameters and processing techniques. The following interactions are particularly important (CORSPERS, 1977):

(a) The interrelationships of soil moisture, soil type, roughness and vegetation cover.
(b) The effective depth of the soil moisture radar measurement and its relationship to the soil moisture profile.
(c) The extent to which microwave radiation penetrates vegetation. This is important not only for soil moisture measurements but also for mapping topographic features flooding and snow cover, and for classifying.
(d) The effect of the complex liquid and crystalline structure of ice and snow on the return radar signal and its penetration depth.

In all sensor applications, the usefulness of the data will depend on the choice of frequenty or frequencies and the polarization(s) used. A proof-of-concept mission (SIR-B) will use multiple depression angles of two frequencies (C band and X band), and several polarization combinations. A comparison of SIR-B with SIR-A is given in Table 6.3 (JPL, 1982). Data from such a sensor will allow extensive analysis to optimize the information content of several simultaneously-received signals, both radar alone and radar in conjunction with sensors.

6.3 CONCLUSION

Two factors that influence the earth's carbon cycle, deforestation and desertification, can be monitored by remote sensing. Accuracy in identifying changes in vegetation cover may be enhanced by proper selection of spectral bands; narrower bands will be available in the future as linear array

Table 6.3 Comparative summary of SAR parameters for SIR-A and SIR-B

Parameters	SIR-A	SIR-B
Orbital altitude	260 km	225 km
Orbital inclination	38 deg	57 deg
Frequency	1.28 GHz	1.28 GHz
Polarization	HH	HH
Look angle(s)	47 deg	15–60 deg
Swath width	50 km	20–50 km
Peak power	1 kW	1 kW
Antenna dimensions (m)	9.4×2.16	10.7×2.16
Antenna gain	33.6 dB	33.0 dB
Bandwidth	6 MHz	12 MHz
Azimuth resolution	40 m (6 look)	25 m (4 look)
Range resolution	40 m	58–17 m
Optical data collection	8 h	8 h
Digital data collection	0	25 h
Digital link capability	N/A	46 Mbits/s

technology improves. However, precision analysis will be required if the system relies on mapping, and the use of computer techniques will facilitate the geographic referencing required. LANDSAT-4 provides improved sensing spatially and spectrally. Synthetic aperture radar promises to provide all-weather sensing, although operational spacecraft are still in the future.

6.4 ACKNOWLEDGEMENT

The above work was performed under NASA Contract NAS 7-100 to the Jet Propulsion Laboratory, California Institute of Technology, while the author was on assignment to NASA headquarters.

6.5 REFERENCES AND BIBLIOGRAPHY

Abrams, M. J., Ashley, R. P., Rowan, L. C., Goetz, A. F. H., and Kahle, A. B. (1977) Mapping of hydrothermal alteration in the cuprite mining district, Nevada, using aircraft scanner images for the spectral region 0.46 to 2.36 μm. *Geology*, **5**(12), 713–718.

Barker, G. R. (1975) Operational considerations for the application of remotely sensed forest data from LANDSAT or other airborne platforms. *NASA, 1975b.*

Billingsley, F. C. (1975) Noise considerations in digital image processing hardware. In: Huang, T. S. (ed), *Topics in Applied Physics*, Vol. 6. Springer-Verlag, New York.

Billingsley, F. C. (1981) Modeling misregistration and related effects on multispectral classification. *Publication 81-6*, Jet Propulsion Laboratory, Pasadena, California. (Condensed version in *Photogrammetric Engineering and Remote Sensing*, **48**(3), 421–430.)

Calspan Corp. (1976) Image processing study. *NASA Contract Report NAS5-20366*, Goddard Space Flight Center.

Chafaris, G. A. (1978) Definition of the major applications and requirements relative to overlaying SEASAT-A SAR and LANDSAT MSS data sets. General Electric Co. In: Darrel Williams (Technical Monitor), *Report contract NAS5-23412*, to GSFC.

Collins, W. (1978) Remote sensing of crop type and maturity. *Photogrammetric Engineering and Remote Sensing*, **44**(1), 43–55.

Colvocoresses, A. P. (1979) Multispectral linear arrays as an alternative to LANDSAT-D. *Photogrammetric Engineering and Remote Sensing*, **45**(1), 67–69.

CORSPERS (1976) Resource and environmental surveys from space with the thematic mapper in the 1980's. *NRC/CORSPERS-76/1*, NTIS. National Research Council, Washington, DC.

CORSPERS (1977) Microwave remote sensing from space for earth resource surveys. *NRC/CORSPERS-77/1*, *NTIS*. National Research Council, Washington, DC.

DelGrande, N. K. (1975) An advanced airborne infrared method for evaluating geothermal resources. *Proceedings of the 2nd U.N. Symposium on the Development and Use of Geothermal Resources*. San Francisco, California.

Ecosystems International, Inc. (1977) Crop Spectra Workshop. *NASA Contract NASW 3024*. Sterling, Virginia.

Evans, W. E. (1974) Marking ERTS images with a small reflector. *Photogrammetric Engineering*, **40**(6), 665–672.

Friedman, H. D. (1965) On the expected error in the probability of misclassification. *Proceedings of IEEE 53*, 658.

Gausman, H. W., Escoban, D. E., Everitt, J. H., Richardson, A. J., and Rodriguez, R. R. (1978) Distinguishing succulent plants from crop and woody plants. *Photogrammetric Engineering and Remote Sensing*, **44**(4), 487–491.

Haralick, R. M. (1975) A resolution preserving textural transform for images. *IEEE Proceedings of Conference on Computer Graphics, Pattern Recognition and Data Structure*, 51–61.

Hilbert, E. E. (1977) Cluster coding algorithm—a joint clustering/data compression concept. *Publication 77-43*, Jet Propulsion Laboratory, Pasadena, California.

Hsu, S. (1978) The Mahalanobis classifier with the generalized inverse approach for automated data processing. *Proceedings American Society of Photogrammetry*, 259–264.

Idso, S. B., Jackson, R. D., and Reginato, R. J. (1977) Remote sensing of crop yields. *Science*, **196**, 19–26.

Jayroe, R. R. (1976) Nearest neighbor, bilinear interpolation, and cubic convolution interpolation geographic correction effects on LANDSAT imagery. *NASA Publication TM X-73348*, Marshall Space Flight Center.

JPL (1982) SIR-B Science Plan, JPL Publication 82-78, Jet Propulsion Laboratory, 4800 Oak Grove Drive, Pasadena, CA 91109 (Dec. 1982)

Kauth, R. J., and Thomas, G. S. (1976) The tasselled cap—a graphic description of the spectral-temporal development of agricultural crops as seen by LANDSAT. *Proceedings of Symposium on Machine Processing of Remotely Sensed Data*, 4B-41-51. Purdue University.

Kerr, R. A. (1977) Carbon dioxide and climate: carbon budget still unbalanced. *Science*, **197**, 1352–1353.

Kondratyev, K. Y., Vassilyev, O. B., Grigoryev, A. A., and Ivanian, G. A. (1973) An analysis of the Earth Resources Satellite (ERTS-1) data. *Remote Sensing of the Environment*, **2**, 273–283.

Landgrebe, D. A., Biehl, L. L., and Simmons, W. R. (1977) An empirical study of scanner system parameters. *IEEE Transactions on Geoscience Electronics GE-15(3)*, 120–130.

Lerman, A., Berhard, M., Bolin, B., Delwiche, C. C., Enhalt, D. H., Gessel, S. P., Kester, D. R., Krumbein, W. E., Likens, G. E., MacKenzie, F. T., Reiners, W. A., Stumm, W., Woodwell, G. M., and Zinke, P. J. (1977) Fossil fuel burning: its effects on the biosphere and biogeochemical cycles. *Global Chemical Cycles and Their Alterations by Man*, 275–289. Dahlem Konferenzen, Berlin.

Millard, J. P., Jackson, R. D., Goettelman, R. C., Reginato, R. J., and Idso, S. B. (1978) Crop water-stress assessment using an airborne thermal scanner. *Photogrammetric Engineering and Remote Sensing*, **44**(1), 77–85.

Miller, L. D. (1979) The carbon cycle question and its demand for satellite remote sensing of global forest cover and forest cover depletion rates. *Proceedings of the 13th International Symposium on Remote Sensing of the Environment*. University of Michigan, April 1979.

NASA (1972) Advanced scanners and imaging systems for earth observations. *NASA Publication SP-335*.

NASA (1975a) LANDSAT-D thematic mapper technical working group. Final Report. *Publication JSC-09797*, NASA Johnson Space Center.

NASA (1975b) Earth Resources Survey Symposium, June 1975, Houston, Texas. NASA Document TMX-58168, *Publication JSC-09930*.

NASA (1976) Active Microwave Workshop. *NASA Conference Publication 2030*.

Rabchevsky, G. A. (1977) Temporal and dynamic observations from satellites. *Photogrammetric Engineering and Remote Sensing*, **43**(12), 1515–1518.

Rao, V. R., Brach, E. J., and Mack, A. R. (1978) Crop discriminability in the visible and near infrared regions. *Photogrammetric Engineering and Remote Sensing,* **44**(9), 1179–1184.

Ready, P. J., Wintz, P. A., Whitsitt, S. J., and Landgrebe, D. A. (1971) Effects of compression and random noise on multispectral data. *Proceedings of the 7th Symposium on Remote Sensing of the Environment,* 1321–1343. University of Michigan.

Richardson, A. J., and Wiegand, C. L. (1977a) A table lookup procedure for rapidly mapping vegetation cover and crop development. *Machine Processing of Remotely Sensed Data Symposium,* 284–296. Purdue University.

Richardson, A. J., and Wiegand, C. L. (1977b) Distinguishing vegetation from soil background information. *Photogrammetric Engineering and Remote Sensing,* **43**(12), 1541–1552.

Richardson, A. J., and Wiegand, C. L. (1979) Automating vegetation calendar and table lookup approaches in a classification strategy. Personal communication.

Salmonson, V. V. (1978) LANDSAT-D, a systems overview. *12th International Symposium on Remote Sensing of the Environment, April 1978,* 378–385. University of Michigan.

Thompson, L. L. (1979) Remote sensing using solid state array technology. *Photogrammetric Engineering and Remote Sensing,* **45**(1), 47–55.

Tucker, C. J. (1978a) Are two photographic infrared sensors required? *Photogrammetric and Engineering and Remote Sensing,* **44**(3), 289–295.

Tucker, C. J. (1978b) A comparison of satellite sensor bands for vegetation monitoring. *Photogrammetric Engineering and Remote Sensing,* **44**(11), 1369–1380.

Tucker, C. J. (1980) Radiometric resolution for monitoring vegetation: How many bits are needed? *International Journal of Remote Sensing,* **1**(3), 241–254.

Tucker, C. J., and Maxwell, E. L. (1976) Sensor design for monitoring vegetation canopies. *Photogrammetric Engineering and Remote Sensing,* **42**(11), 1399–1410.

Tucker, C. J., and Miller, L. D. (1977) Soil spectra contributions to grass canopy spectral reflectance. *Photogrammetric Engineering and Remote Sensing,* **43**(6), 721–726.

University of Michigan (1979) Papers and poster papers presented at the *13th International Sympoisum on Remote Sensing of the Environment.*

Vincent, R. K. (1973) A thermal infrared ratio imaging method for mapping compositional variations among silicate rock types. Doctoral thesis, University of Michigan.

Westin, F. C., and Lemme, G. D. (1978) LANDSAT spectral signatures: studies with soil associations and vegetation. *Photogrammetric Engineering and Remote Sensing,* **44**(3), 315–325.

Weszka, J., and Rosenfeld, A. (1975) A comparative study of texture measures for terrain classification. *IEEE Proceedings of Conference on Computer Graphics, Pattern Recognition and Data Structure,* 62–64.

Wiegand, C. L., Richardson, A. J., and Kanemasu, E. T. (1979) Leaf area index estimates for wheat from LANDSAT and their implications for evapotranspiration and crop modeling. *Agronomy Journal,* **71**, 336–342.

The Role of Terrestrial Vegetation in the Global Carbon Cycle:
Measurement by Remote Sensing
Edited by G. M. Woodwell
© 1984 SCOPE. Published by John Wiley & Sons Ltd

CHAPTER 7
Coupling Remotely Sensed Data to Ground Observations

A. B. PARK
Natural Resources Consulting Services, 606 Shore Acres Road, Arnold, Maryland, USA

ABSTRACT

Remote sensing is increasingly employed by those interested in mapping landscapes. Data received by the remote sensor, in this case LANDSAT, are not perfectly correlated with the land. No map is 100 per cent accurate. Sequential imagery and use of ratios of one band to another have been shown to improve accuracy greatly. An example is offered of an approach that might be used to examine changes in vegetation globally.

7.1 INTRODUCTION

Colwell (1956) published the first reports of an important non-military use of colour infra-red film, he found that black stem rust of wheat and yellow dwarf virus disease ih oats could be detected before symptoms were visible to the unaided eye. One might have expected this report to have caused rejoicing among phytopathologists, but initial pleasure yielded to a decade of debate on why the technique worked. No satisfactory explanation based on a biotic mechanism was offered. The scientific community was reluctant to accept without further evidence that the correlation between the image and the disease constituted effective prediction. However, the popular press and semi-technical journals published many articles on the identification of disease using remotely sensed imagery.

More than 10 years later Gausman and others (1969, 1971) explained the phenomenon. They showed that differences in intracellular water pressure and intercellular air spaces were responsible for near infra-red reflectance; rather than identifying disease *per se*, Colwell's film had detected loss of cell turgor. The coupling of the imagery to the disease required another, more difficult step based on additional information. The correlation between the observation and the cause in this case proved to be correct, but such inferences based on data from remote sensing may be misleading.

After 1956 there was a gradual evolution in remote sensing from qualitative recognition to what might be called an enumeration phase of measuring features of the landscape. One can think of the former phase as exploration, the latter as mapping.

7.2   THE MAPPING PHASE

However, mapping is not a simple transformation of data from remote sensing. Clients of photogrammetry frequently expect, unrealistically, a perfect correlation between the data derived from remote sensing and ground observations. Several factors complicate this correlation:

(1) Measurements made remotely are by definition the response of the target through the atmosphere.
(2) Precision is limited by the accuracy of the instrument and its resolving power.
(3) In most situations it is not possible to resolve pure target material, i.e., one can not see plants in a garden without seeing the soil.
(4) The time of day and season influence the target and the radiant energy directed twoards the sensor.

These general limitations govern remote sensing analyses as follows:

1. The difference between the identification of features and the discrimination of features is of paramount importance. Given the current state of technology, the identification of features is not performed by an objective analysis of the measurements made by the sensor. Rather, the identification process occurs either because the shape of the feature, or the 'colour' of the feature, or both, are unique and within the experience of the image analyst. (The term 'colour' includes the spectral response of the feature whether or not it is in the visible portion of the electromagnetic spectrum.) Discrimination, on the other hand, occurs solely because the 'colour' of different features is, in fact, different. For this case the analyst may not and usually does not know the identity of the discriminant. Identification occurs only after a field survey. Conventionally, this survey is done by the analyst but in some cases it may be provided as collateral data in the form of reports and/or photographic interpretation keys.

2. Once the identification process has been accepted, the matter of accuracy remains. If the parameters involved in the acquisition phase are examined, it becomes clear that 100 per cent accuracy is never achievable. Although there are some features for which remote sensing technology can approach 100 per cent (i.e., bare soil, green vegetation, water) within these classes there will be a

finite range of values for each class and a probability for error associated with each range of values.

These comments apply to all maps. No map is 100 per cent accurate, and no inventory or survey or even sample taken during the course of a survey is free of error. Remote sensing has one important advantage over every other survey technique; it can and does enumerate the entire population. This enumeration merely reduces the errors, it does not eliminate them. The limitation is caused by the heterogeneous nature of those features that we consider to be members of the same population. There is, for example, a large but finite range of sizes and shapes and colours of those species of trees we class as deciduous or coniferous. Even if we only include a single species in our class, for example, wheat, the statement is still true. In fact, if one examines a taxonomic text from the statistical point of view, one finds that the descriptive approach follows the familiar bell-shaped Gaussian curve.

Fortunately, the data behave according to conventional statistical principles and the accuracy of the survey or inventory with and without field work can be determined. In addition, one can employ statistical methods to calculate the improvements one can derive from additional sampling (i.e., field verification) versus the cost of the improvement. It must be remembered that the potential accuracy can only be estimated from experience with data drawn from similar populations. For example, if an analysis of LANDSAT data has been performed in one portion of a large ecosystem (the grasslands) one can reasonably expect to get similar results in another part of the same ecosystem providing the analysis is performed at the same time.

7.3   INTERPRETATION OF IMAGERY

Two simple rules help in interpreting remotely sensed imagery: (1) work with ecosystems, and (2) recognize the importance of the date of the image and the possibility of comparing images from different times. The utility of these rules becomes apparent from a consideration of the radiometric properties of the LANDSAT-D Thematic Mapper (Table 7.1) and the response of the mapper to vegetation (Table 7.2).

Different species of plants vary little in responses and the possibilities for separating species in one image seem limited. But when images at different times are available, considerably more information can be used. Band ratios when compared over time give information descriptive of the age, condition, health of the plant, or of canopy density relative to the background from soils, become useful in identifying species or communities. Interpretation of the data in the context of a phenological model is thus critical in the process of interpretation.

For most crop plants, three observations at different times are usually required to identify the species; a fourth may be required to make some of the

Table 7.1 Radiometer characteristics for the thematic mapper

Band	Spectral width (μm)	Dynamic range (mw/cm²-ster)	Low level input (mw/cm²-ster)	S/N′
1	0.45–0.52	0–1.00	0.28	32
2	0.52–0.60	0–2.33	0.24	35
3	0.63–0.69	0–1.35	0.13	26
4	0.76–0.90	0–3.00	0.16	32
5	1.55–1.75	0–0.60	0.08	13
6	2.08–2.35	0–0.43	0.03	5
7	10.40–12.50	260 °K–320 °K	300 °K	0.5 °K (Net D)

Absolute chennel accuracy < 10 per cent of full scale.
Band to band relative accuracy < 2 per cent of full scale.
Channel to channel accuracy < 0.25 per cent rms of specified noise levels.
S/N′ = signal to noise ratio.

Table 7.2 Thematic mapper spectral and radiometric characteristics

Band	Wavelength	NEΔρ	Basic primary rationale for vegetation
TM 1	0.45–0.52	0.008	Sensitivity to chlorophyll and carotenoid concentrations
TM 2	0.52–0.60	0.005	Slight sensitivity to chlorophyll plus green region characteristics
TM 3	0.63–0.69	0.005	Sensitivity to chlorophyll
TM 4	0.76–0.90	0.005	Sensitivity to vegetational density or biomass
TM 5	1.55–1.75	0.01	Sensitivity to water in plant leaves
TM 6	2.08–2.35	0.024	Sensitivity to water in plant leaves
TM 7	10.4–12.5	0.5 °K	Thermal properties

NEΔρ = noise equivalent reflectance difference.

more difficult separations. In the LACIE project (see Erickson, this volume), which developed techniques for predicting world wheat production, there was an absolute requirement of a series of three or four observations over the season.

Recent studies by Park have shown that information on the stage of growth of crops can be derived from repetitive observations using ratios of one band to another. The bands are plotted as vectors (Figure 7.1(a)–(f)). Results showed that the ratios of bands 4/5 compared with 6/5 crossed very shortly after emergence of the crop. At harvest there was another cross over. Careful use of the ratios with knowledge of the phenology enabled identification of the following stages of growth in corn on the satellite imagery:

(1) Preplanting (bare soil)
(2) Emergence

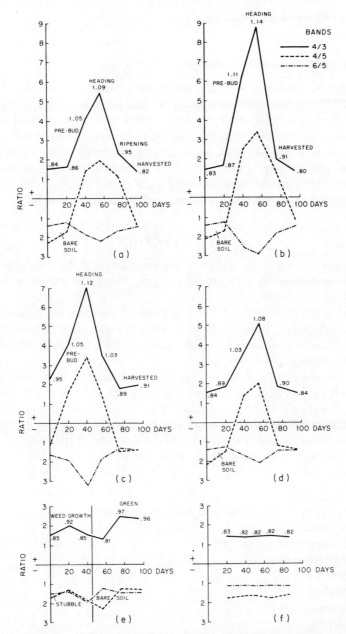

Figure 7.1 Ratios of airborne spectrometer measurements, simulating multispectral scanner (MSS) bands from LANDSAT 4 (thematic mapper), plotted as vectors over time: (a) spring wheat; (b) barley; (c) alfalfa; (d) oats; (e) corn; (f) bare soil ((a)–(e) Hand County, South Dakota; (f) Williston County, North Dakota). These ratios may be used to identify stages of plant development

(3) Booting or prebud (preheading)
(4) Heading
(5) Ripening
(6) Harvest (post-harvest)

Knowledge of phenology is vitally important in interpretation, and use of the knowledge requires sequential imagery. The most important advantage that LANDSAT and other satellite imagery offers over other types of remote sensing is that global measurements can be made repetitively. Most of the experience with vegetation has been developed in attempts to estimate agricultural productivity. In this work the satellite imagery is used in conjunction with a model that estimates yield of the crop on the basis of meteorological data, information from the satellite and other sources.

7.4   SITE SELECTION

The first consideration in the design of the programme is the selection of the area to be represented by each cell of the grid. There are several important criteria.

First, the scale must be determined. If national statistics are required, the grid cells can be larger than would be possible if regional or provincial data are sought. Reducing the size of the cells increases the number of cells and the homogeneity among cells.

Second, the complexity of the terrain must be estimated. If the crop grows under a wide range of conditions and is widespread, cell size can be large. If the crop grows only under a narrow range of conditions or is managed intensively, as is the case with rice, cell size must be reduced. The decision on cell size can be made readily from a LANDSAT mosaic of the country.

After a decision is made on size, each cell must contain the following descriptors:

(1) Area of the crop
(2) Number of crops harvested and rotation practices
(3) Crop calendar
    (a) Planting or transplanting
    (b) Emergence
    (c) Flowering
    (d) Heading
    (e) Ripening
    (f) Harvest
(4) Crop varieties
(5) Potential yield
        Because it is possible to model the potential yield in more than one way, it is useful to store historical yield data as follows:

   (a) Minimum
   (b) Maximum
   (c) Mean
 (6) Cultivation practices
 (7) Fertilizer
 (8) Plant pests
      Commonly, a visual report will be used to start a review of remotely
      sensed imagery to:
      (a) Confirm the event
      (b) Determine the area affected
      (c) Note the presence of additional susceptible species
      (d) Confirm the stage of growth predicted by the yield model, and
      (e) Estimate the effects
 (9) Soil
(10) Meteorological data
(11) Terrain model

7.5 APPLICATION OF LANDSAT DATA

The design of the data base for a field crop is shown in Figure 7.2 in graphic
form. This model, with its global data base and emphasis on crops, is an
example of one potential approach to monitoring the role of vegetation in the
carbon cycle. Important work is also possible at the other end of the scale; the
desertification process. A number of researchers are using LANDSAT data to
study local effects. Robinove and Chavez (1978) have developed an important
process of coupling vegetated areas of the earth with adequate spatial resolution.
With Robinove and Chavez's approach, atmospheric and sun elevation
corrections are applied with no ancillary data required. Trends in the change of
albedo (indications of the darkening or lightening of the terrain) form the basis of
a monitoring scheme in arid and semi-arid regions. Such variations can be the
result of changes in the density or type of vegetation, changes in the area of bare
ground, or changes in the distribution of wind-blown material. Slight snow or
cloud cover in a full scene does not significantly affect the mean albedo of large
areas. The ease of the method recommends it for widespread use and further
experimentation.
   Robinove and Chavez's work is a good example of the modelling required
to couple LANDSAT data to a very slowly changing surface phenomenon.
The coupling involved the calibration of the albedo measurements in terms of
vegetation trends. Their work is also a good example of a principle now
accepted by the various disciplines; ground observation programmes suffer
without LANDSAT data to show investigators where to make surface
observations and measurements.

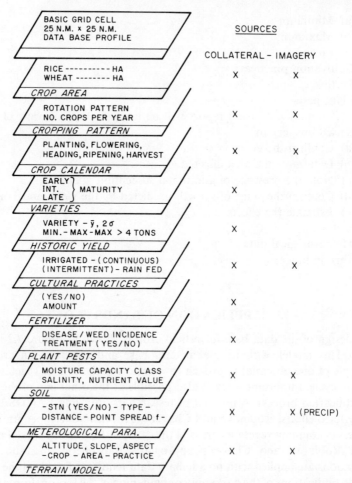

Figure 7.2 Data-base designed for on-line support of phenology and yield models of field crops

7.6 SUMMARY

The coupling of remote sensing data with ground observations is accomplished by an appropriately trained scientist. This is particularly true for the enumerative mapping function, in which the image is transformed into a map of the terrain. Field trips may or may not be required. As one proceeds from mapping to monitoring dynamic features of the terrain, such as vegetation and water resources, extensive use is made of models. In most cases the pixel radiance values become coefficients in an equation which measures some attribute of plant condition other than the area it covers. Most applications which are operationally oriented are supported by a geo-referenced data base. Here, however, the data are transformed into resource management decisions.

7.7 REFERENCES

Colwell, R. N. (1956) Determining the prevalence of certain cereal crop diseases by means of aerial photography. *Hilgardia*, **26**, 223–286.

Gausman, H. W., Allen, W. A., and Cardenas, R. (1969) Reflectance of cotton leaves and their structure. *Remote Sensing of Environment*, **1**, 19–22.

Gausman, H. W., Allen, W. A., Wiegand, C. L., Escobar, D. E., Rodriquez, R. R., and Richardson, A. J. (1971b) The leaf mesophylls of twenty crops, their eight spectra, and optical and geometrical parameters. *SW C Res. Report 423*. 88 pp.

Keegan, H. J., Schleter, J. C., Hall, Jr., W. A., and Haas, G. M. (1956) Spectrophotometric and colorimetric study of diseased and rust resisting cereal crops. *National Bureau of Standards Report 4591*.

Park, A. B. (1978) Multispectral temporal analyses. *IR&D Report*. 54 pp.

Robinove, C. J., and Chavez, Jr., P. S. (1978) LANDSAT albedo monitoring method for an arid region. AAAS International Arid Lands Conference on Plant Resources, Lubbock, Texas.

Tucker, C. J. (1978) A comparison of satellite sensor bands for vegetation monitoring. *Photogrammetric Engineering and Remote Sensing*, **44**(11), 1369–1380.

Wiegand, C. L., Gausman, H. W., and Allen, W. A. (1972) Physiological factors and optical parameters as bases of vegetation discrimination and stress analysis. *Proceedings, Seminar on Operational Remote Sensing*, American Society of Photogrammetry. 341 pp.

The Role of Terrestrial Vegetation in the Global Carbon Cycle:
Measurement by Remote Sensing
Edited by G. M. Woodwell
© 1984 SCOPE. Published by John Wiley & Sons Ltd

CHAPTER 8

The LACIE Experiment in Satellite Aided Monitoring of Global Crop Production

J. D. ERICKSON
NASA Johnson Space Center, Houston, Texas, USA

ABSTRACT

The Large Area Crop Inventory Experiment (LACIE) demonstrated that improved accuracy in USDA predictions of wheat production can be achieved for the US Great Plains by the use of satellite imagery. LACIE experimenters also used their technique to predict with great accuracy the size of the 1977 Soviet wheat crop six weeks prior to harvest. This paper discusses the experiment as a potential model for other programmes designed to measure globally other terrestrial plant communities by remote sensing from satellites.

8.1 INTRODUCTION

The Large Area Crop Inventory Experiment (LACIE) was carried out by the National Aeronautics and Space Administration (NASA), the US Department of Agriculture (USDA) and the National Oceanic and Atmospheric Administration (NOAA). The objective was to develop and test a method for estimating production of wheat worldwide. The experiment was intended:

(1) To demonstrate an economically important use of repetitive, multispectral, remote sensing from space;
(2) To test the capability of the LANDSAT, together with climatological, meteorological and conventional data sources, to estimate in advance the size of an important world food crop;
(3) To validate techniques that can provide timely estimates of crop production.

The basic approach used in the experiment was to combine estimates of the land area planted in wheat with estimates of yield per unit area. Estimates of area were derived from LANDSAT data on selected segments of land, estimates of yield were obtained from models which relied on weather data from the World Meteorological Organization. The experiment included

191

The role of terrestrial vegetation in the global carbon cycle

computer processing of data and the use of mathematical models to obtain information in a timely manner.

In August 1977 the experiment produced what later proved to be an accurate estimate of the shortfall in the Soviet spring wheat crop. This observation was well before definitive information about the crop was released by the USSR. In addition, analysis of spring and winter wheat production in the Soviet Union during two other crop years resulted in estimates that supported the experiment's goals for performance. The success of the LACIE experiment was reinforced by accurate estimates of production in the US winter wheat region for three crop years.

The experiment was less successful in predicting Canadian wheat production, but the reasons are well understood. They were that effective field size in Canada was often very close to the resolution limits of the LANDSAT, and that spring wheat is difficult to distinguish from certain other crops.

LACIE resulted in the development of a technique for estimating overall wheat production on the basis of area and yield estimates, a technique of acceptable accuracy for estimating crop area without the use of ground data, and a technique of acceptable accuracy for estimating crop yields.

Refinements of the procedures for analysing LANDSAT data can further improve the satellite's accuracy in identifying land area planted in wheat. Yield models may be improved by utilizing LANDSAT data together with weather data to better define the crop's response to natural conditions. Models that estimate the crop's stage of development can also be improved to provide data that will help to distinguish wheat from similar crops (such as barley) and thus also lead to improved forecasts.

LACIE was a timely response to an identified national need and to a specific need. It was the culmination of more than a decade of research and development, it assembled a special array of people and equipment, and it was rigorously tested on a large scale. LACIE's encouraging results have led to further efforts to determine the requirements of the USDA and other users and to extend the capability to other important questions.

The experiment was initiated in 1974 to assimilate remote sensing by satellite and its associated communications techniques into an experimental system and to use that system to make production estimates of important crops. Wheat was selected for the experiment both because of its economic importance and because it would fit well with the evolution of space technology. Wheat is grown on huge areas of the United States and the Soviet Union as well as on smaller plots in India and China. It is grown in some part of the world on every day of the year. As well as being one of the least complex crops from an agricultural standpoint, wheat was also one of the crops most amenable to remote sensing. It seemed likely that techniques developed to predict wheat production more accurately might be adaptable to other crops.

Agricultural production is highly variable, since it is dependent on the complicated interactions of weather, soils, technology, and other factors. The agricultural outlook can and usually does change as these ingredients are altered, either by natural forces or as a result of human decisions.

To forecast agricultural production accurately, it is vital to associate the correct weather with the actual crop area being affected. When the effects of the weather are so severe as to remove an area from agricultural production, the area must be correctly measured. Therefore, an agricultural information system must monitor not only the total area harvested but also the proportion of the area rendered worthless from an agricultural standpoint by weather extremes.

8.1.1 The Background of LACIE

In 1960 the Agricultural Board of the National Research Council recommended the formation of a committee to investigate the potential ability of aerial surveys to monitor agricultural conditions over large geographic areas. An interdisciplinary group of scientists was then selected to serve on the Committee on Remote Sensing for Agricultural Purposes, and by late 1962 the Committee had designed experiments to examine the feasibility of using multispectral remote sensing to monitor crop production. This step was followed in 1965 by the establishment of an organized research programme, by the USDA and NASA. The programme led, from the creation of the first multispectral scanner and computer recognition of wheat from multispectral measurements collected by aircraft in 1966, to (1) the identification of the spectral bands and other design characteristics of the first Earth Resources Technology Satellite (ERTS) in 1967; (2) a simulation of ERTS data from the SO-65 multispectral photographic data taken by Apollo IX in 1969; (3) the successful launching of ERTS in 1972 and (4) feasibility investigations in 1972 and 1973 which demonstrated the potential ability of the ERTS system to monitor important crops.

Investigations into the relationships between weather and crop yield have been an agricultural research interest of long standing. The availability of high-speed computers and worldwide weather data in recent decades has allowed much more extensive statistical analysis of those relationships. Some researchers have studied the responses of individual plants to the weather, while others have investigated the subject on a larger scale to determine the relationship between average yield and normal climatic conditions in specific regions. Several of these studies, undertaken at Iowa State University around 1970, investigated key relationships among yield, agricultural technology, and climate in the major grain-producing area of the United States. On the basis of that work NOAA initiated a study in 1973 to evaluate the likelihood of drought in the Midwest and the possible effects of such a drought on grain production.

LACIE was a logical next step. By then, a technological apparatus consisting of earth observation satellites, environmental satellites, communications links, computer processing equipment and mathematical models had been created. In LACIE these elements were assembled into a system capable of large-scale monitoring of global wheat production.

8.1.2   Roles of the Federal Agencies

Each of the three federal agencies participating in LACIE brought specific expertise and experience to the planning and implementation of the experiment. Most of the LACIE tasks required the integrated efforts of at least two of the three agencies; however, various lead responsibilities were assigned. Figure 8.1 illustrates the participation of the three agencies.

8.1.3   Role of Universities and Industry

Researchers from universities and industry played a key role in the experiment through the development of improved techniques that were evaluated in later phases of LACIE and through participation in technical review sessions held

Figure 8.1   The three federal agencies participating in LACIE

periodically throughout the experiment. In addition key industries were, through contracts from the agencies, vital to the implementation and operation of the experiment.

8.2 THE LACIE EXPERIMENT

8.2.1 Objectives of the Experiment

The objectives of LACIE included the following:

(1) To demonstrate an economically important application of repetitive, multispectral, remote sensing from space;
(2) To test the capability of the LANDSAT, together with climatological, meteorological and conventional data sources, to estimate the production of an important world food crop;
(3) To validate techniques that can provide timely estimates of crop production;
(4) To provide estimates of the area planted in wheat, to provide estimates of wheat yield and to combine these area and yield estimates to estimate total production;
(5) To develop data processing and delivery techniques so that a selected sample could be made available for analysis no later than 14 days after acquisition of the data;
(6) To develop a LACIE system design that with a minimum of redesign and conversion, could be used to develop an operational system within USDA;
(7) To monitor and assess crop progress from a surface data base and evaluate the model potential for yield and surface data.

Ancillary goal-oriented activities included:

(1) Periodic crop assessment from planting through harvest;
(2) Support for a research and development (R and D) programme to improve methodology and performance;
(3) An objective test and evaluation programme to quantify the results of R and D.

To maintain the experimental nature of LACIE, it was decided that crop assessment reports would be prepared on a monthly basis during the crop season and mailed to USDA the day before the official USDA monthly report was released. The goal was to make periodic estimates of production that would be, on the average, within ± 10 per cent of actual production 90 per cent of the time (referred to as the 90/90 criterion). An additional goal was to establish the accuracy of these estimates early in the season (the first quarter of the crop cycle) and continue their accuracy through the harvest period. The three agencies agreed that achievement of the 90/90 criterion would be an improvement over information available from conventional data sources.

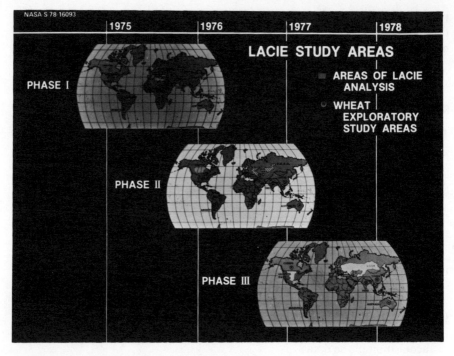

Figure 8.2   LACIE study areas

8.2.2   Scope of the Experiment

The LACIE experiment was designed to monitor production in selected wheat-producing regions of the world. The experiment extended over three global crop seasons and was designed to include up to eight regions (Figure 8.2). The early phases of the experiment concentrated primarily on the nine-state wheat region in the US Great Plains, where current information about wheat production and the components of production was available to permit quantitative evaluation of LACIE operations. As the experiment progressed, it expanded to include the monitoring of wheat production in two other major wheat-producing countries, Canada and the USSR. This expansion also included exploratory studies for monitoring wheat production in India, China, Australia, Argentina and Brazil (Figure 8.2). In addition, management decisions by USDA resulted in the incorporation of a USDA User System within the USDA–LACIE effort.

Phase I of LACIE (global crop year 1974–75) focused on the integration of components into a system to estimate the proportion of the major producing regions planted in wheat, and on the development and feasibility testing of yield and production estimation systems. At the end of the season, a report on

LACIE estimates of wheat and small grains production in various areas of the US Great Plains was prepared.

In Phase II (global crop year 1975–76) the technique, as modified during Phase I, was evaluated for its accuracy in monitoring wheat production on the US Great Plains, in Canada, and in 'indicator regions' in the USSR. Monthly reports of area, yield and production estimates of wheat for these regions were prepared.

8.2.3 Technical Approach

The technical approach to LACIE (Figure 8.3) was to estimate wheat production on a region-by-region basis. Both the area planted in wheat and the yield were estimated for local areas and aggregated to regional and country levels to determine production. Maximum use was made of computer-aided analysis in order to provide the most timely estimates possible. The estimates were made throughout the crop season, and evaluations were conducted to verify the accuracy of the LACIE technique and to identify technical problems.

Estimates of the area in wheat were from LANDSAT 2 multispectral scanner (MSS) data acquired for land segments of 5 × 6 nautical miles. The use of LANDSAT full-frame imagery allowed the taking of samples from agricultural areas only and meant that an analysis of only two per cent of the sample was sufficient. Sampling error was less than two per cent. The techniques of statistical pattern recognition employed in LACIE were designed to take advantage of the changing spectral response of crop types over time. Thus, LANDSAT data were acquired throughout the crop season, screened to determine cloud cover, and registered to previous acquisitions. The sample segments were then extracted in a digital format. Trained analysts then labelled a small part (less than one per cent) of each sample segment as either wheat or not wheat. In general, the analysts were not able to distinguish wheat from other small grains in a reliable manner. Therefore, the labelling was generally of small grains, and historically derived ratios were applied to the small grains estimates to estimate wheat. (A procedure for direct identification of spring wheat, based on subtle differences in crop stages and appearances was tested late in Phase III.) The labelling was based on the appearance of wheat as observed over time on digital film imagery of each segment and on graphical plots indicating the response in each of the spectral channels. Because the spectral appearance of the crop is a strong indication of its stage of growth, models were developed for estimating the growth stage, based on local weather data. The analysts were also provided with summaries of seasonal weather and of local cropping practices for each region.

Wheat yield was estimated using statistical regression models based on recorded wheat yields and weather in each region. These regression models

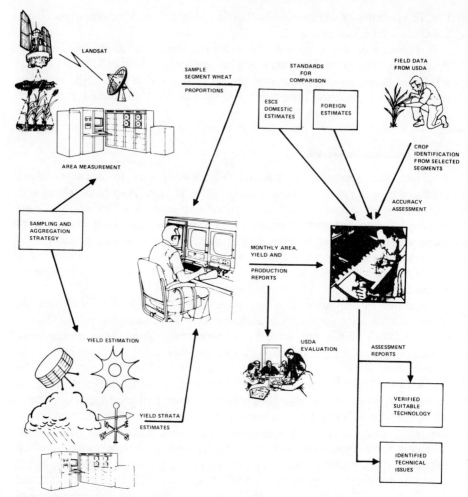

Figure 8.3 LACIE technical approach

forecast yield for fairly broad geographic regions, using calendar month values of average air temperature and cumulative precipitation. Meteorological data for these models (as well as the growth stage models) and weather summaries for the Great Plains were obtained primarily from surface stations of the National Weather Service, the Federal Aviation Agency and the Department of Defense. In foreign areas the data were collected by each country's weather service and were made available to LACIE by means of network of the World Meteorological Organization. Imagery for both foreign and domestic areas that was obtained by satellites was used to refine the precipitation analyses, which were based on cloud patterns. Models were developed to make yield estimates early in the season, throughout the growing season and at harvest. Estimates of winter wheat yields in the northern hemisphere began in December and were updated until harvest in June or July. Estimates of spring wheat yields began as early as March and were revised monthly through August or September. Thus, assessments of potential yields were made almost from the time the plants emerged from the ground.

8.3   RESULTS OF THE EXPERIMENT

Perhaps the most important results of the LACIE experiment were of a technique to provide dramatically improved information on wheat production in important global regions and the demonstration that the technique could respond in a timely manner to large weather-induced changes in production. The most graphic example of this capability involved the LACIE prediction of the 1977 Soviet wheat crop.

8.3.1   The Phase III Results from the USSR

As shown in Figure 8.4, in January 1977 the Soviet Union set a goal of 120×10^6 metric tons (MT) for its wheat crop that year. In August 1977 the LACIE experiments made an initial forecast that total Soviet wheat

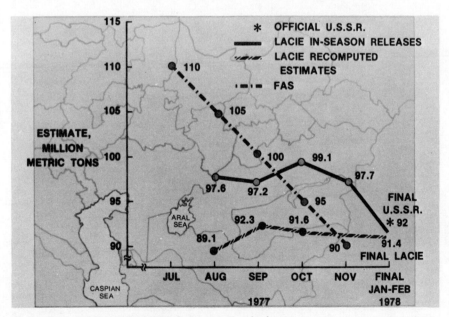

Figure 8.4   LACIE 1977 USSR forecasts: Total wheat

production would amount to 97.6×10^6 MT, or 20 per cent below the January goal of the Russians. This was only six per cent above the final Soviet figure of 92.0×10^6 MT. The final LACIE estimate of 91.4×10^6 MT differed from the final Soviet figure by about one per cent.

In comparison with the accuracy and timeliness of wheat crop information emanating from the USSR, these results showed an important advance in forecasting ability. Prior to the LACIE experiment, it was necessary to rely heavily on statistics and reports released by foreign countries themselves. Apart from questions about the reliability of such information, the major problem is its timeliness. The Russians, for example, release only a planning figure for grain production early in the year and a post-harvest estimate of total grain production in early November. Actual statistics are not released until the January or February following the harvest. The wheat production forecasts of the Foreign Agricultural Service (FAS) of USDA (shown in Figure 8.4) were based to a large extent on Soviet reports and to a lesser extent on reports from foreign agricultural attachés. The LACIE-recomputed estimates in Figure 8.4 resulted from a smoothly functioning operational system that could produce estimates of wheat production 30 days following the acquisition of data by the LANDSAT.

Figure 8.5 shows the separate winter and spring wheat estimates that constitute the totals in Figure 8.4. The May and June forecasts of winter wheat

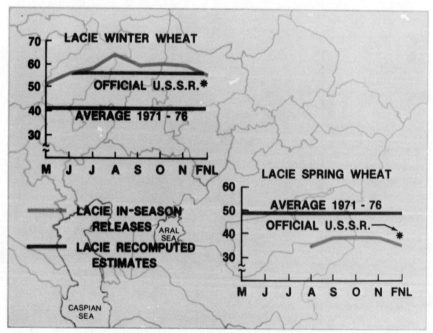

Figure 8.5   LACIE 1977 USSR forecasts. Winter and spring wheat

were for a normal to above-normal crop. The increase from May to June was known (because of LACIE forecasting experience in the US) to be the result of the steadily increasing visibility of the wheat crop to the LANDSAT. However, the continued increases in the July and August forecasts of winter wheat could not be justified, either on the basis of improving detectability or improving weather. Thus, alerted to technical problems, LACIE analysts initiated efforts to isolate the source of these reports of further increases. (Spring wheat estimates were unaffected by the problem and stabilized, as expected, following the August forecast.)

By November the problem in the forecast of winter wheat was discovered to be the result of a faulty LANDSAT data acquisition order, which led to the loss of key early season data on about 20 per cent of the sample segments of winter wheat. For these segments, only spring data were available, and LANDSAT analysts could not differentiate between winter wheat and small spring grains, such as barley, which had already become detectable. Even though the LACIE forecasts were accurate despite the implementation problems, 'recomputed estimates' were generated in December of 1977 to simulate the performance of a system without the data order problem. To generate the recomputed estimates, wheat output for the winter wheat areas affected by the faulty data orders was computed by utilizing the original estimate for those areas as an estimate of total small grain production, which was then reduced to a winter wheat figure using historic ratios of winter wheat to total small grains. In addition, a problem arising from using data 45 to 60 days old in current reports was eliminated by utilizing data acquired up to 30 days prior to the reporting date.

The clues to the production shortfall in the spring wheat region of the USSR came early in the season, when weather conditions started on an unfavourable note. The average air temperature for May and June was as much as 55 per cent above normal throughout the region. As a result, the wheat needed a greater amount of moisture than usual. (It is evident from Figure 8.6 that the abnormally high temperatures were widespread.)

But rainfall during the same period was below normal in many of the crop regions, as shown in Figure 8.7. The above-normal need for moisture, combined with the below-normal supply, clearly indicated that a serious problem was developing. (Figure 8.8 shows where the deviations from the normal supply–demand relationship were most pronounced.)

Since differences between precipitation and potential evapotranspiration were used in LACIE models to represent the relative soil moisture available to the crop, it was natural to expect a significant detrimental effect in the eastern and southern crop regions. The drought conditions were clearly observable in the LANDSAT data, and LACIE yield models responded accurately by reducing yield estimates in the affected regions. (As Figure 8.9 shows, the reduction in many cases was 50 per cent below normal.) In response

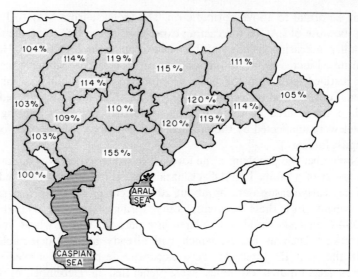

Figure 8.6 USSR spring wheat region: Per cent of normal for May–June air temperature (monthly average, °C)

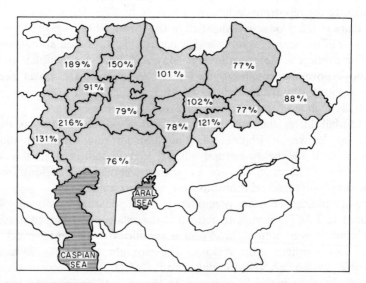

Figure 8.7 USSR spring wheat region: Per cent of normal for May–June monthly precipitation (mm)

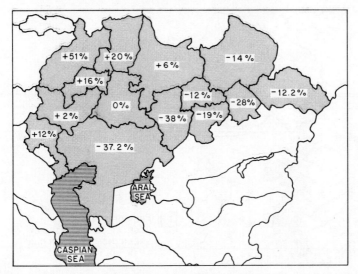

Figure 8.8 USSR spring wheat region: Per cent deviations from normal May–June monthly precipitation minus potential evapotranspiration (PET) (mm)

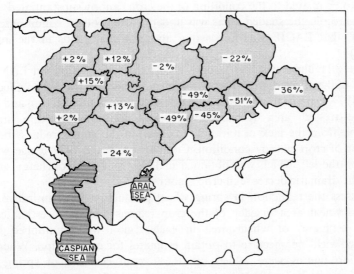

Figure 8.9 USSR spring wheat region: Per cent deviations from trend yields (quintals per hectare) (10^5 g/ha)

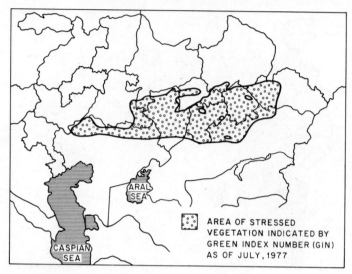

Figure 8.10   USSR stressed vegetation

to the high temperatures in April, before the spring season commenced LACIE yield models showed a probable loss of nearly 2×10^5 g/ha. The continued drought reduced below the normal figure of 11.5×10^5 g/ha. It can be seen in Figure 8.10 that these drought conditions were also quite evident from the LANDSAT data. In this figure, radiometric measurements from LANDSAT known to be related to the condition of the crop canopy substantiated the fact that the crop in the shaded areas was under severe drought conditions. (Note, however, that LACIE was forecasting above-normal yields in the northern regions.)

Figure 8.11 illustrates the drought effects that were visible on LANDSAT imagery of the affected area. The two-segment images on the right, collected on July 4, 1977, were from a normal moisture area (Omsk Oblast, top) and from a moisture-stressed area (Kokchetav Oblast, bottom). Moisture stress is detectable from the lack of darkness (redness) in the image, redness being an indicator of crop canopy condition. Comparison of the 1977 image with the image on the left, which was obtained in 1976 from the Kokchetav segment, shows the dramatic decrease in crop vigour in 1977.

To assess the reduction in spring wheat production in quantitative terms, the total wheat area in each of the crop regions had to be estimated. The LACIE estimates of wheat area in each region were multiplied by the forecasted yield per hectare to obtain estimates for each region. When these production figures were added up, the overall estimate of spring wheat production was 36.3×10^6 MT, a deviation of about 21 per cent below normal.

The LACIE yield and acreage estimates have been empirically tested by a

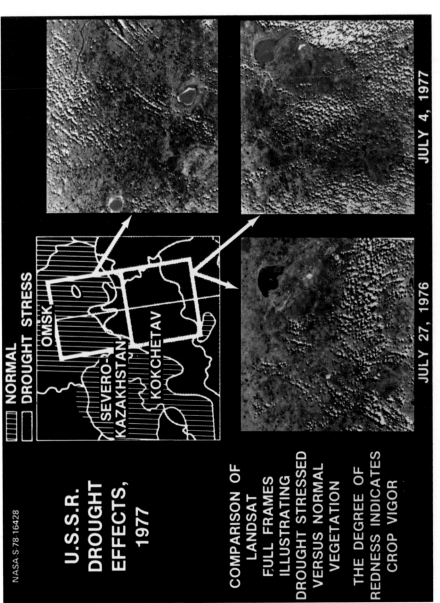

Figure 8.11   USSR drought effects visible on LANDSAT 1977

fairly large number of 'performance experiments'. The LANDSAT-derived estimates of acreage have been evaluated through comparisons with independent ground observations and USDA estimates for the United States, and foreign and USDA estimates for Canada and the USSR.

The LACIE yield models, whose performance is much more sensitive to weather than are the acreage estimates, have been evaluated for the same countries with the aid of more than 10 years of historic data. While these years and regions are quite different from each other and represent a reasonable sample of the potential conditions to be encountered in a global survey, these empirical estimates can be viewed with increasing confidence as their number increases over the years.

Later in this paper it is stated that in some cases the LACIE technique achieved the 90/90 criterion and that in other cases it did not. These statements, based on certain statistical assumptions generally believed to be quite valid, represent inferences drawn from the performance tests described above.

How much confidence can be placed in these statements? LACIE used a standard, well-accepted approach to data that has not contradicted the 90/90 hypothesis except in those cases noted. The experimental data do not contradict the 90/90 for United States winter and USSR total wheat. While a lack of contradiction of this hypothesis implies that the LACIE technology may be satisfying 90/90 in a region, increased confidence can only be gained through additional replications over a number of years.

8.3.2 Phase III Results in the United States

Phase III in the United States further substantiated the conclusion that the technical modifications incorporated into the experiment during Phase II worked well. Overall, the Phase III results (Figure 8.12) showed significant improvement over those of Phase II, as winter wheat estimates were indicative of 90/90 accuracy. In addition, there was significant improvement during Phase III in the ability to estimate the spring wheat area. This reduced the difference between the LACIE estimate and the estimate of the Economics, Statistics and Cooperative Services (ESCS) on wheat area to less than one per cent, compared to a Phase II difference of 13 per cent. Unlike the Phase I and II results, the Phase III estimates of yield were significantly below those of the ESCS and were not supportive of the 90/90 criterion. However, the yield estimates in combination with the improved area estimates resulted in production estimates which differed from those of ESCS by less than 10 per cent. Statistical tests indicated that the Phase III estimates of United States wheat production were probably accurate enough to achieve the 90/90 criterion. The Phase III estimates of area, yield and production for the United States Great Plains region are shown in Figure 8.12. The yield estimates

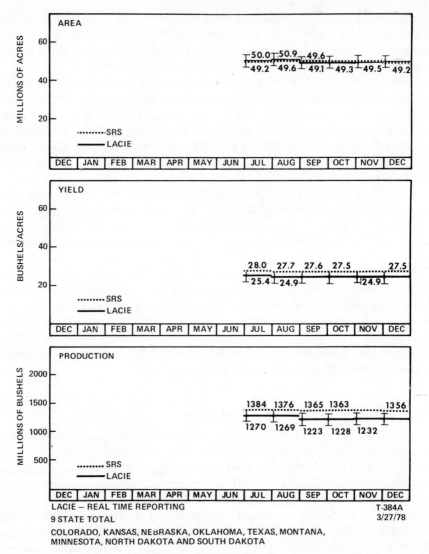

Figure 8.12 Monthly comparison of LACIE and SRS estimates. US Great Plains (9 state total), Phase III

shown are not the results derived from individual yield models, they were derived by dividing total production by total acreage. Even though the final estimate of yield was made in September, the derived value changed slightly as later LANDSAT data were used to refine estimates of area.

More extensive evaluations of the yield models over a 10-year period indicated performance consistent with the 90/90 criterion except in years with

Table 8.1 Results of an evaluation of the LACIE Phase III US yield models on 10 years of independent test data

Year	SRS, bu/acre	LACIE bu/acre	Error	Within tolerance
1967	21.6	22.5	+0.9	Yes
1968	26.0	24.6	−1.4	Yes
1969	28.4	29.4	+1.0	Yes
1970	28.2	26.6	−1.6	Yes
1971	30.8	27.9	−2.9	No
1972	29.3	29.1	−0.2	Yes
1973	30.8	30.6	−0.2	Yes
1974	23.8	28.4	+4.6	No
1975	26.8	27.3	+0.5	Yes
1976	26.4	27.1	+0.7	Yes
1977 [a]	27.5	24.9	−2.6	

Mean error = 0.1 bu/a
RMSE = 1.90 bu/a

[a] Phase III results.

extreme agricultural or meteorological conditions. Table 8.1 lists the results of a test of Phase III yield models using data for the years 1967 to 1976. The models were developed from data for the 45 years prior to each of the test years. A non-parametric statistical test employed to analyse this data did not reject the 90/90 hypothesis. However, had the models exceeded the tolerance bounds in at least one more year (as they appear to have done in 1977), the 90/90 hypothesis might have been rejected. In addition, the root mean square error (RMSE) of 1.9 bushels per acre (bu/a) was larger than desirable for determining whether the 90/90 criterion had been achieved. It should be noted, however, that 1974 was a dry year in the Great Plains, and wheat yields were very poor. The LACIE yield models failed to respond to this deviation and overestimated yield by 4.6 bu/a. Omitting 1974, the RMSE would drop to 1.3 bu/a, which is not significantly different from the figure required for a 90/90 estimator. Thus, it appears that the yield models may satisfy the 90/90 criterion in years where there are no extreme fluctuations in yield.

Also, evaluated in Phase III were the LACIE models of the stages of wheat growth. These models, which were of key importance to the analysis of LANDSAT data, predicted the growth stage of wheat given maximum and minimum daily air temperatures. Generally, the Phase III evaluation of these models indicated that improvements were required, particularly the development of a model to predict the planting date. Given accurate data on the planting date, however, the models seemed to perform adequately. Improved growth stage prediction models are also the key to improved yield models.

Phase III testing of alternate sampling strategies in the United States and the USSR indicated that substantial cost savings could be realized through their use. Three improved strategies will permit accurate estimates to be made with significantly reduced amounts of data.

The results achieved during Phase III in predicting production on strip fallow (small field) areas in the spring wheat regions of the United States showed significant improvement but still exhibited a tendency to underestimate the area of small grains. Figure 8.13 shows the experimental estimates in comparison with ESCS estimates. Figure 8.14 compares LACIE estimates of wheat area percentages, at the segment level, with observations of actual percentages made by human observers on the ground ('ground truth'). These 'ground truth' data were prepared independently of, and after, the Phase III estimates from LANDSAT data were produced. This comparison shows the improvement in Phase III results.

The actual time required to analyse a LANDSAT segment, manually select training fields, compute training statistics, and process the nearly 23 000 elements of the segment was reduced from 10 to 12 hours during Phase I to 6 to 8 hours during Phase II and 2 to 4 hours in Phase III. It was also concluded that the experiment showed that the timeliness goal of 14 days could be realized in the future.

The geographically dispersed nature of the LACIE data processing system led to long 'in-work' times (from 30 to 50 days) for segments of LANDSAT data because of the many manual steps required and the fact that the experiment was conducted, for the most part, on a one-shift, five-days-a-week basis. The actual time during which a segment was undergoing processing, however, was within the revised goal of 14 days from acquisition to availability for aggregation, since actual 'contact time' was two to four hours per segment and computer processing time was around five to eight minutes per segment.

8.3.3 Phase II Results in the United States, USSR and Canada

While the Phase III results were very encouraging, they were by no means the whole story. The experimental results obtained in the United States during the three years of LACIE, and in the Soviet Union during Phase II, also substantiate the Phase III results in the USSR. The estimates of US and Canadian spring wheat defined the geographical regions for which improvements in remote sensing technology were needed.

An evaluation of Phase II results indicated that the technique worked well for winter wheat in the United States and for both winter and spring wheat in the USSR. Difficulty was encountered, however, in reliably differentiating spring wheat from other spring small grains, primarily spring barley, in the spring wheat regions of the United States and Canada. The strip fallow fields in these regions were another complicating factor, since their widths were very

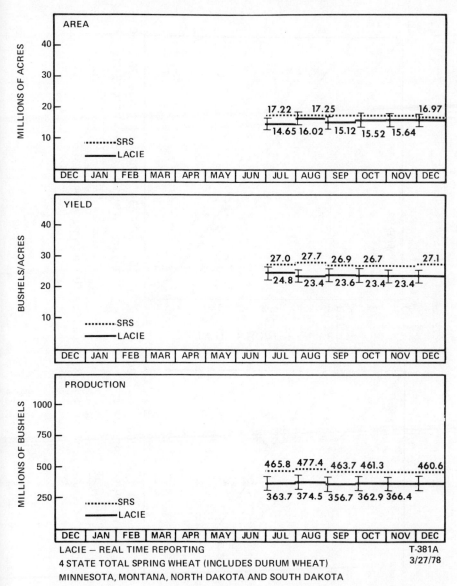

Figure 8.13 Monthly comparison of LACIE and SRS estimates. US Great Plains (4 state total), Phase III

close to LANDSAT resolution limits. Figure 8.15 shows how field size and shape were problems in some areas. On the left part of Figure 8.15 is an aerial photograph and segment of the strip/fallow region of the United States. Note the prevalence of very long and narrow fields, a result of moisture-conserving

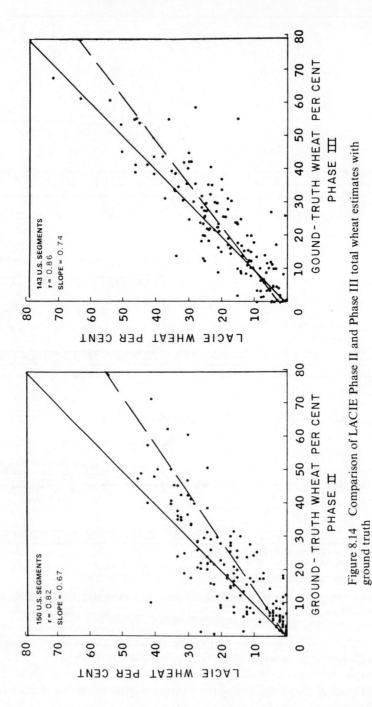

Figure 8.14   Comparison of LACIE Phase II and Phase III total wheat estimates with ground truth

strip/fallow practices. Similar practices are common in Canadian spring wheat areas.

These factors led to significant Phase II underestimates of 29 and 26 per cent for the spring wheat areas of the United States and Canada, respectively. In the spring wheat regions of the USSR, where field sizes are considerably larger and the ratios of wheat to small grains are more stable than in the US and Canadian regions, the Phase II estimates were in reasonable agreement with crude estimates based on official Soviet statistics. In 1977 there were other indications, such as estimates of the coefficient of variation of the LACIE estimates, that the LACIE estimates were of 90/90 quality. Replications are required to verify the achievement of the 90/90 criterion, however. The final LACIE estimate was within one per cent of the Soviet figure. Most encouraging was the accuracy of the estimates made early in the growing season.

The decision to expand the region to be inventoried in the USSR was prompted by the lack of actual production information for the USSR indicator regions and thus the absence of a reliable estimate of the bias of the LACIE estimates. It was also decided to reduce coverage in Canada to 30 segments, where Canadian investigators could use ground observations in an intensified evaluation of the problems of distinguishing spring wheat from other small grains in small fields.

8.3.4   Exploratory Foreign Investigations

Exploratory investigations in Argentina, Australia, Brazil, China and India provided insight into the technical problems of estimating production in other countries. These investigations included the development of yield models, analysis of sample segments and collection of LANDSAT, meteorological and agronomic data. Aggregated estimates of area, yield and production were not attempted.

8.3.4.1   Australia

LANDSAT data collected over Australia indicated field sizes and multi-temporal signatures similar to those of the US Great Plains and the USSR. Yield models have been developed for five states in Australia. A test of these models, using 10 years of independent test data, indicated they would support the 90/90 criterion. A model to predict the stages of crop growth has also been developed, but difficulties have been encountered in using it because wheat varieties in Australia differ from those grown in the United States, where the model was developed. The model was designed for winter wheat with a dormancy period; Australian wheat does not go into dormancy.

Figure 8.15 Resolution of LANDSAT images

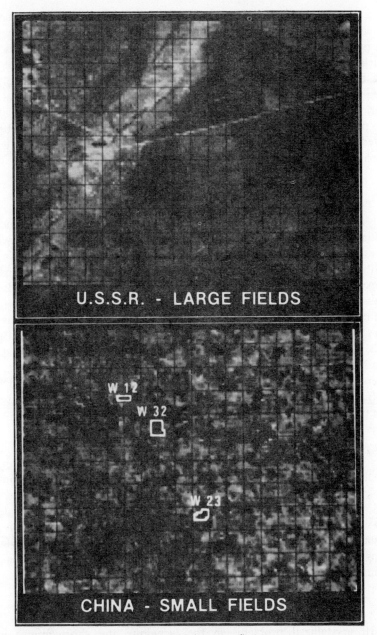

Figure 8.15 (*continued*)

8.3.4.2 India

The average field in India is smaller than the LANDSAT MSS resolution element. However, the fields tend to be adjacent to each other and may be less of a problem than small strip fields in the United States and Canada. Yield models have been developed for 15 states in India, and exploratory segments have been analysed. Although not tested operationally, these models were tested using historic data, and the test indicated they would support the 90/90 criterion. Models to predict the stages of growth were evaluated and showed very poor results. Many of these can be attributed to differences between United States and Indian wheat strains. Indian wheat does not go into dormancy and has a shorter growth cycle.

8.3.4.3 Argentina and Brazil

Analysis of LANDSAT data indicated that wheat fields in the older and more populated areas of Argentina are similar in size to those in Kansas, while in the less populated frontier areas they are similar in size to wheat fields in the USSR. Ancillary data for Argentina and Brazil were extremely limited, thus hampering both interpretative analysis and the creation of yield models. Yield regression models were developed for five provinces in Argentina and for one state in Brazil, but the quality of the data for building these models was lower than the quality of the data available for equivalent United States areas. Tests of the yield models, using more than 10 years of independent test data, indicated that the models would not support the 90/90 criterion. In general, the crop signatures were typical of those encountered in the United States. LANDSAT data on Brazilian wheat regions showed that cloud cover was more frequent there than in the United States wheat region.

8.3.4.4 China

China, like India, has extremely small wheat fields in its densely populated areas, but field sizes are comparable to those in the United States in the newly developed spring wheat region. Historical data upon which to develop necessary ancillary data could not be found. This lack of information about the stages of crop growth and about crops hard to distinguish from wheat by means of satellite observations meant that less confidence could be placed on the analysis of Chinese wheat production.

8.3.5 Technological Problems Requiring Further Attention

LACIE crystallized problems in the technique and shortcomings in our understanding of certain phenomena. Problems in need of special attention include the following:

(1) The need to develop yield models that are based on daily or weekly, rather than monthly, averages of temperature and precipitation, and that more closely simulate the critical biological functions of the wheat plant and its interaction with the external environment.
(2) The need to develop techniques to deal more effectively with the spatial information in LANDSAT data and to improve the accuracy of area estimates in regions where a high percentage of the fields have effective sizes close to the resolution limit of LANDSAT. In addition, further investigation of the improvements resulting from the increased resolution power of LANDSAT-D, as well as the spatial resolution requirements for future LANDSAT satellites, is necessary.
(3) The need for better understanding of the distinctly different characteristics of wheat grown in tropical regions.
(4) The need for better quantification of the effects of cloud cover on the acquisition of LANDSAT data at critical periods in the crop season, particularly in more humid environments, such as the United States corn-belt.
(5) The trade-offs between the need to shorten the time from data acquisition to reporting and the cost of obtaining a quicker response. Although it is possible to reduce this time span, doing so may require substantial additional costs.

8.4 CONCLUSIONS

On the basis of the results obtained by LACIE during three crop years, we conclude that:

(1) It is now possible to estimate wheat production successfully in geographical regions whose characteristics are similar to those of the Soviet Union's wheat areas and the winter wheat area of the United States.
(2) Significant improvements in our ability to estimate wheat production in these and other regions can be expected in the near future through additional applied research.
(3) The remote sensing and weather effects modelling techniques developed in LACIE may be applicable to other major crops and regions of the world.

In addition, several lessons were learned about the planning, management and implementation of programmes to develop improved crop monitoring and estimation techniques:

(1) Research, development and evaluation require several years of testing with large amounts of data to verify technological issues, due to the wide range of variability in the factors that contribute to the outcome.
(2) A comprehensive effort to assess accuracy is vital. Substantial amounts of

ground data from domestic 'yardstick' or test regions are essential to understanding experimental results as well as to the identification and correction of deficiencies in the programme. LACIE included this type of assessment.

(3) A research and development programme involving diverse scientific disciplines, focused on technical issues that arise from a project like LACIE, stimulates applied research and provides an improved understanding of the programme in the academic and industrial communities.

(4) The periodic use of a peer review, in which critical methodological issues are subjected to the scrutiny of reviewers from university, government and industry provides essential feedback.

(5) Much was learned about the capabilities of the LANDSAT, together with other data sources, to estimate wheat production. Most importantly, the needs for higher spatial resolution, additional spectral bands, and increased coverage to observe smaller fields and to distinguish wheat from other crops, were identified. LANDSAT-D will provide a data source to support solution of technical problems related to these needs.

8.5 OUTLOOK

As a result of (1) USDA's continued interest in exploiting this technique to provide improved information about crop production throughout the world, (2) the success achieved thus far with wheat and (3) the identification by LACIE of technical issues requiring further investigation, the US Secretary of Agriculture called for the creation of a multiagency programme to develop improved uses of aerospace technology for agricultural purposes. The AgRISTARS programme focuses on the following:

(a) early warning of environmental or technological changes that may affect the production or the quality of renewable resources;
(b) improved commodity production forecasts;
(c) land-use classification and measurements;
(d) renewable resources inventory and assessment;
(e) land productivity estimates;
(f) assessment of conservation practices; and
(g) pollution detection and evaluation.

While all seven of these are of major importance to USDA, the first two express the Department's urgent need for better and more timely objective information on world crop conditions and expected production. The agencies that participated in LACIE planned activities for the early 1980s that would build on the LACIE experiment and address the broader needs of USDA.

The results of the LACIE experiment also provide a partial basis for judging the feasibility of measuring global vegetation and its changes. LACIE provided numerous research results that are also relevant to forest and rangeland measurements, to the use of imaging radar, and to measuring insolation and solar temperatures by means of meteorological satellites. Generally speaking, these research areas show promise but have not been evaluated on a global scale to demonstrate that reliable technology has been achieved.

8.6 REFERENCES

Document of Plenary Papers (1978) LACIE Symposium. Houston, Texas.

LACIE Executive Summary (1978) Houston, Texas.

LACIE Management Plan—Baseline Document (1976) (LACIE C00167, JSC-11334). Houston, Texas.

LACIE Phase I Evaluation Report (1976) (LACIE-00318, JSC-11663). Houston, Texas.

LACIE Phase II Evaluation Report (1977) (LACIE-00453, JSC-11694). Houston, Texas.

LACIE Project Plan—Baseline Document (1975) (LACIE-C00605, JSC-09857). Houston, Texas.

MacDonald, R. B., Hall, F. G., and Erb, R. B. (1975) *Proceedings of Second Symposium on Machine Processing of Remotely Sensed Data, 1B-1–1B-23*. McGillem, C. D. (ed). Institute of Electrical and Electronics Engineers, Inc., New York.

MacDonald, R. B., Hall, F. G., and Erb, R. B. (1975) The Large Area Crop Inventory Experiment (LACIE)—An assessment after one year of operation. In: *Proceedings of 10th International Symposium on Remote Sensing of Environment*, Vol. 1, 17–37. Environmental Research Institute of Michigan, Ann Arbor, Michigan.

Memorandum of Understanding Among the US Department of Agriculture, The National Aeronautics and Space Administration and the National Oceanic and Atmospheric Administration for the Experimental Large Scale Crop Inventory Demonstration. (1974) LACIE files, Washington, DC.

National Research Council (1977) World Food and Nutrition Study: The Potential Contributions of Research Steering Committee, Commission on International Relations. National Academy of Sciences, Washington, DC.

Proceedings of the 1974 Lyndon B. Johnson Space Center Wheat-Yield Conference (1975) (LACIE-T00407). Houston, Texas.

SECTION V

Conclusion

There are many different approaches to the use of remote sensing in measuring changes in the vegetation of the earth, each of them promising. Several have been discussed in this book. Recent research, summarized in the following pages, suggests that one of these approaches is considerably more promising than others.

The Role of Terrestrial Vegetation in the Global Carbon Cycle:
Measurement by Remote Sensing
Edited by G. M. Woodwell
© 1984 SCOPE. Published by John Wiley & Sons Ltd

CHAPTER 9

Measurement of Changes in the Vegetation of the Earth by Satellite Imagery

G. M. WOODWELL,* J. E. HOBBIE,* R. A. HOUGHTON,* J. M. MELILLO,*
B. MOORE,† A. B. PARK,‡ B. J. PETERSON,* AND G. R. SHAVER*
* The Ecosystems Center, Marine Biological Laboratory, Woods Hole,
Massachusetts, USA
† Complex Systems Center, University of New Hampshire, Durham, New
Hampshire, USA
‡ Natural Resources Consulting Services, 606 Shore Acres Road, Arnold,
Maryland, USA

ABSTRACT

Substantial progress has been made recently in use of LANDSAT imagery for
measuring the changes in the vegetation of the earth that must be recognized to resolve
questions about the global carbon cycle. Progress has required the use of a
mathematical model designed around the principles of succession. The data for the
model are of two types: basic ecological information about succession, including
especially rates of storage of carbon, and information on changes in land use such as
rates of transformation of forests to non-forests and non-forests to forests. These latter
data are most easily obtained by comparison of images taken sequentially. Primary
reliance is placed on detection of change, not on classification and inventory. An
example is given of how such data are used to determine the net exchange of carbon for
a region such as the State of Maine. A global sampling plan designed to reduce the
current uncertainty in the rate of deforestation from sixfold to less than 25 per cent
would require a sample of less than one-tenth of the 12 000 scenes required to cover the
total land surface of the earth.

9.1 INTRODUCTION

Pictures of the entire earth are now available as imagery from space. No
corner of the earth has escaped the probing eye of LANDSAT or any of
several other satellites, each arranged to record portions of the electromagnetic
spectrum. And the imagery can be repeated, daily in some instances, almost
twice monthly in others. The techniques are sufficiently versatile that new
methods of investigation are possible using existing data to make further
improvements in analysis of details of the earth's surface.

Despite this achievement, supported by hundreds of millions of dollars from the public trust, there is no clearly defined, simple method for obtaining global estimates of such attributes of the earth as the area of forests and changes in that area year-by-year. The need for the information is well-defined. It has been recorded in this book as information basic to resolving a series of uncertainties with respect to the global carbon cycle. But it might have been set forth with equal power as a need for information about changes in albedo, or rates of transpiration, or rates of runoff of water from once forested lands, or about areas in agriculture, or simply as an inventory of wood or energy or the potential for fixing energy biotically, all fundamental to life.

It is possible now to prepare a map of the vegetation of the entire earth from LANDSAT imagery. The problem is cost. Despite the millions available for the hardware, the money available for applications, even to fill fundamental human needs, is small. How can the specific information needed at any moment be extracted with elegance and economy from the wealth of information, equipment and opportunity available now?

Answers to such questions are never simple. Some of the complexity is indicated in the foregoing analyses: the enormous amount of scholarship that has been devoted to the interpretation and mapping of vegetation; the uncertainties inherent in interpreting the status of the carbon retained in soils; the bewildering intricacies of the film and other equipment used for remote sensing by aircraft and satellite; and for the problems of knowing, not assuming on tenuous evidence, what the image shows.

But there is simplicity and clarity emerging, too. Objectives are being narrowed. A new array of global problems is emerging with the CO_2 problem seen as central. The progress is both in narrowing objectives and in limiting the approaches to those that are most promising.

The existence of continuous coverage of the earth with several different types of imagery leads to the assumption that many comprehensive questions can be answered by some type of map. The promise is great. The four spectral bands recorded by LANDSAT offer a wide array of possibilities for developing and testing 'signatures' and it is easy to believe that unique signatures can be found for virtually any objective. This apparent versatility encourages the natural tendency toward classification and mapping, but it is misleading in that it encourages an expectation of great detail at small expense. Satellite imagery does not replace aerial photography. The need for critical choices, a winnowing of objectives, persists. Nowhere is this need for simplicity more clear than in the application addressed in this book.

9.2 DETECTION OF CHANGES IN VEGETATION

The summary of those aspects of the CO_2 problem presented above (Chapter 1), drawn from more detailed analyses (Woodwell *et al.*, 1978; Woodwell, 1978;

Moore *et al.*, 1981; Houghton *et al.*, 1983; Woodwell *et al.*, 1983a,b), suggests that the largest advance in determining the role of the biota in the global carbon cycle at present can be made by improved information on rates of deforestation, especially in the tropics. There is, of course, the possibility that in certain areas, such as the northeastern United States, the abandonment of agriculture has led to extensive reforestation (Woodwell *et al.*, 1978; Houghton *et al.*, 1983); any system for estimating net release from the biota must be capable of detecting reforestation as well. Changes such as these, from forest to non-forest or the reverse, commonly involve large changes in reflectance that are easily detected; the number of classes is limited to those that are most easily identified. Subtleties of elaborate classification schemes can be avoided and emphasis placed on detecting a small number of transitions, each of which can be recognized accurately over large areas.

There is a further advantage in avoiding classification in so far as possible. A complex classification or mapping system opens the possibility of errors due solely to the classification, not to errors in measurement of change. A misclassification of a forest that had not changed over the period considered, for example, would appear as a change in the inventory. There is no need to measure areas that are not changing with respect to their carbon content. Attention can be focused exclusively on areas that have been changed recently.

Despite the simplification, the avoidance of complex classification of vegetation, and the focus on change itself, identification of the types of change may require several years if the only source of information is satellite imagery (Figure 9.1). The forest to non-forest change is recognizable immediately, but

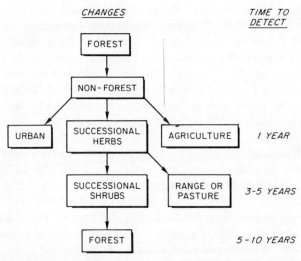

Figure 9.1 The use of satellite imagery for detection of types of changes may require successive images over several years

clear evidence that agriculture has replaced forest or that forest has been allowed to recover may not be available from satellite imagery for several years.

Such information, once available, must be tabulated to be useful in interpretation of the carbon cycle. The tabulation requires a model of some sort. The model used at present is based on succession. All sources of information are used to provide detailed analyses of the stocks of carbon in the vegetation and soils regionally and globally (Moore *et al.*, 1981; Houghton, *et al.*, 1983). The system is well suited for use with information from LANDSAT, especially if the data are derived as outlined here from an emphasis on detection of change as opposed to much more elaborate classification of the vegetation.

These two processes, the development of new data on changes in the vegetation of the earth, and development of a model capable of providing summaries of the net flux of carbon between the land and the atmosphere, are linked; neither can proceed with precision alone. A model that has been developed for this purpose is described in the pages that follow, and one application of it using satellite data from Maine is outlined. The experience suggests that the uncertainty surrounding current estimates of global deforestation can be reduced inexpensively to between one-half and one-tenth of the current level by a satellite-based survey.

9.3   A GLOBAL MODEL

Interpretation of how changes in vegetation and soils affect the amount of CO_2 requires knowledge of natural vegetation, knowledge of patterns of succession, and an understanding of patterns of disturbance over recent centuries. If the interpretation is to be global, a careful accounting will be needed to allow for the complexity of the transformations and to tabulate them over time for each of the various types of vegetation in different regions. It is necessary to know how the carbon in vegetation and soils changes as a result of disturbance and how much land is exposed to different kinds of disturbance each year.

To meet these needs, to provide a systematic and objective appraisal, and to keep a record of the changes over years to a century or so, a simple bookkeeping model has been developed (Moore *et al.*, 1981; Houghton *et al.*, 1983; Woodwell *et al.*, 1983a). The central principle was plant succession. Succession occurs following the harvesting of forest or the abandonment of agricultural or grazing land (Figure 9.2). At the time of harvest the stature of a forest is reduced by some fractional amount, and succession starts. At the same time there is a transfer of additional organic debris into the humus, insolation increases at the surface of the soil, soil temperature rises, and decay is

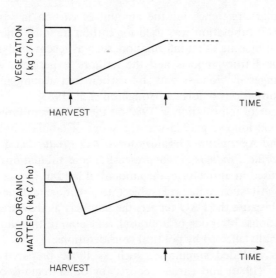

Figure 9.2   Idealized curves describing the changes per unit area in biomass and soil carbon that take place during the regrowth of a harvested forest

stimulated. Enhanced decay continues for several years, reducing the organic content of the soil below the amount that originally occurred in the forest (Covington, 1981). Changes following conversion of forest to agriculture are discussed in this volume in the chapters by Buringh and Schlesinger. When the carbon accumulating in the successional community more than balances the loss of carbon that occurs through respiration, the organic content of the soil begins to rise. Transfers of carbon into the harvested products, where the mean residence time varies between 1 and 1000 years, also were incorporated into the model.

Ten geographic regions were recognized in the model and twelve plant communities (ecosystems) were accommodated. The curves were modified for different initial stocks of carbon, different periods of succession, different degrees of recovery prior to further harvest, and different intensities of harvest. The transfer of carbon from the plant community into soils was modified to correspond to the magnitude of the transfer and the rate of decay. Details of the construction of the model were drawn from the literature and the experience of the authors (Moore *et al.*, 1981; Houghton *et al.*, 1983).

Two kinds of data are needed to operate such a model. One is the amount of carbon held in the vegetation and soils before and after disturbance; the other is the rate of disturbance. These data must be estimated if they are not available from field studies.

Information on the amount of carbon held in the vegetation and soils of the earth is limited. The tabulation prepared by Whittaker and Likens (1973) was used as the primary reference for the amount of carbon in vegetation and Schlesinger's (1977) tabulation was used for carbon in soils. These data were modified to accommodate other information, such as recent analyses of Brown and Lugo (1980). Earlier reports and the authors' experience were used to describe the changes in biomass and soil carbon that accompany succession and other transformations of forests (Houghton *et al.*, 1983).

The main source of information in 1980 on the changes in forest areas over time for the period following 1945 was the series of tabulations compiled by the UN Food and Agriculture Organization (FAO *Production Yearbooks* and *Yearbooks of Forest Products*, 1946–present). These tabulations include the amount of land used in agriculture, the amount of land in forests, and sales of forest products. All such data are subject to error, sometimes substantial (Persson, 1974), because the FAO reports the numbers as they are provided by the reporting nations. Methods of appraisal vary among the nations, and the data are occasionally affected by political considerations.

Other sources included summaries such as those prepared for tropical forests by Myers (1980) and earlier reports for other regions of the world. These records do not cover the entire world, however, and a comprehensive set of contemporary data obtained independently through objective methods is needed.

9.4 APPLICATION OF SATELLITE IMAGERY

The greatest simplicity in use of satellite imagery for the purpose outlined here appears to be in a system designed around detection of change using paired images. Various approaches have been used previously (Robinson, 1979; Riordan, 1982). One of the simplest is based on subtracting the spectral information of one LANDSAT scene from the information of a second scene of another date. The two images are brought into precise coincidence, one over the other, a process called registration. One image is subtracted from the other, both images being in digital form within the computer. Areas that have the same position and spectral information are nullified; areas that are different become part of the new, third image. The greatest difficulty is in obtaining accurate registration of one image on the other. Various techniques have been developed for facilitating the process (Woodwell *et al.*, 1983a).

The advantages of the approach based on detection of change as opposed to any system of classification are several:

(a) The paired image change detection technique avoids most classification.
(b) It eliminates errors associated with classification, a major potential source of error in estimating change.
(c) It measures the area of change directly.
(d) It focuses on the large, easily seen and readily measured areas of dramatic change (forest to non-forest, non-forest to forest) while relying

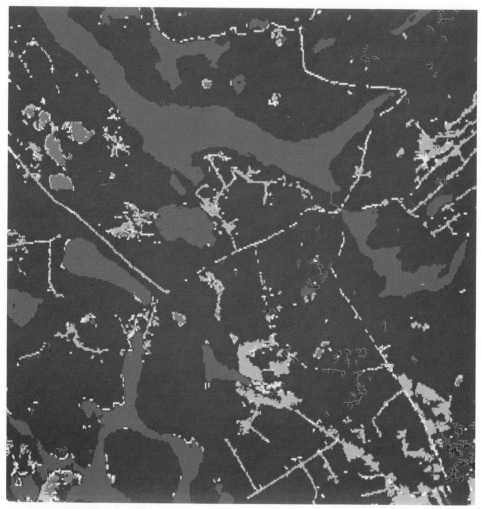

Figure 9.3   Mud Pond subscene, north central Maine. The changes are shown here in yellow for forest to non-forest and in red for non-forest to forest. (For key see Table 9.1)

on a model to compute the more subtle, easily misclassified successional changes in biomass over time.

(e) It estimates efficiently the most uncertain term (clearing and harvest rates) in the terrestrial carbon balance while relying on gound-based data for the terms most difficult to measure via satellite (carbon in biomass and soils of various types of vegetation).

The technique requires, however, certain systematic corrections and much care in application. Registration is usually accurate within ± 1 pixel. A systematic error of one pixel introduces errors into the residual image. The margins of ponds or forest stands may appear as changes. Roads, for instance, are approximately one pixel wide and appear as change if the images are not aligned perfectly. A special algorithm was prepared for this work to eliminate this edge effect (Woodwell *et al.*, 1983a). The correction was based on recognition that:

(a) errors of misregistration occur at boundaries; and

(b) these errors appear as lines one pixel in width.

The algorithm is a filter that eliminates single-pixel lines otherwise recorded as changes.

9.4.1   Implementation of the Method in Maine

The change-detection system has been tested in northern Maine and in the State of Washington using LANDSAT imagery. Figure 9.3 shows one application of the technique in a segment of a scene near Mud Pond in north central Maine. For convenience in presentation the changes have been superimposed on the LANDSAT image, which has been classified as shown (Table 9.1).

The accuracy of such tests was estimated for selected plots in Maine by use of aerial photographs at much larger scale. The same change detection procedure was followed in the photography as for the satellite imagery. The experience revealed that reliance on detection of high-contrast changes, such as

Table 9.1   The Mud Pond subscenes, north central Maine. See Figure 9.3

	Total area (%)	Colour
Water	13.6	Dark blue
Forest	78.4	Green
Bogs	0.6	Light blue
Roads and clear-cuts	1.7	White
Forest to non-forest	4.9	Yellow
Non-forest to forest	0.7	Red

the change from forest to bare ground, may introduce a systematic exaggeration of change beyond the single pixel lines discussed above. The problem arises because pixels on the borders of changes often cover an area that was only partially cleared but the entire pixel was recorded as changed because of the sharp contrast. The bias was corrected by subtracting half of the border pixels from the area recorded as changed. This correction was, of course, most important in areas where there were many small or long-narrow areas that had been changed in some way.

The change detection technique as outlined here was applied to three areas in northern Maine. The areas were selected for convenience in analysis, not because they were representative of Maine. Nonetheless, they provide an example of how data from satellite imagery, once accumulated, can be used with other data and experience to determine details of flux of carbon from changes in forest area.

9.4.2 Estimates of Forest Biomass and the Release of Carbon from Deforestation

The data from three subscenes in northern Maine indicated that 6.88 per cent of the area in forest was cut between September 1972 and July 1977, about 1.38 per cent annually. The total area of forests in Maine, according to Ferguson and Kingsley (1972), was 16 894 000 acres (6.837×10^6 ha) in 1971. If the annual cutting rate obtained from the three subscenes is representative for the state, timber was harvested over 233 000 acres (94 000 ha) annually.

According to US Forest Service studies, the 'growing stock' in Maine forests averaged 1258 cu ft of wood/acre (99 m^3/ha) in 1971 (Ferguson and Kingsley, 1972), or 28 metric tons (MT) C/ha (2.8 kg/m^2) (average density of wood is 700 kg/m^3 according to the FAO, 1979; average carbon content of wood is 45 per cent). Such estimates do not include leaves, branches, bark or roots. The crudest estimate of the net removal of carbon from the forests of Maine in 1975 is the product of area cut (94 000 ha) × biomass (28 MT C/ha), or 2.63×10^{12} g C.

This estimate can be refined considerably. First, growing stock is not equivalent to biomass. Growing stock volume includes only commercial species 5.0 inches of dbh (diameter breast height) or greater, and only the stem wood from stump to a top diameter of 4.0 inches. The rest of the tree above- and below-ground, smaller trees, non-commercial trees and other species may contribute to a biomass 2.7 times greater than growing stock (Johnson and Sharpe, 1982). With this conversion factor the mass of carbon in the trees would be 75.6 MT/ha or 7.5 kg/m^2 and the net release of carbon for the entire State in 1975 was 7.10×10^{12} g, about three times the crude estimate above.

A direct estimate of biomass based on field measurement of 376 samples from Elm Stream Township, taken by H. E. Young (1976) and colleagues, was 7.93 kg/m^2.

A small part of the total biomass is oxidized and released as CO_2 during harvest or immediately after harvest. Most of the wood is used to produce paper, lumber or other products that decay over a longer period. Paper products are usually oxidized in 10 years and lumber in 100 years. The delay in decay makes the records of earlier harvests important in estimating the flux for the current year.

The history of the harvest of forests in Maine from 1860 to 1970 was obtained from Wood (1971), Ferguson and Kingsley (1972), and Smith (1972).* The fraction of the harvest that was paper, lumber, and other products was estimated from reports of the Forest Service.

These data were used with the assumption that the fraction of the harvest in the 1-, 10- and 100-year decay pools has always been as it was in 1971 (6, 65 and 29 per cent) (Ferguson and Kingsley, 1972), to provide an estimate of the net release of carbon from Maine forests in 1975 of 2.36×10^{12} g. This estimate included the flux from wood harvested and burned in 1975 (0.16×10^{12} g C) and the flux from the oxidation of residual products from the last 100 years (2.20×10^{12} g) (Table 9.2).

Much of the original biomass is left on site at the time of harvest. For this analysis we started with a forest biomass of 75 MT C/ha ($2.7 \times$ growing stock of 28 MT C/ha), harvested 28 MT C/ha, and left on site 46 MT C/ha, of which only 5 MT C/ha was assumed alive and 42 MT C/ha was dead. Decay of logging residues from harvests over the past century contributed 4.38×10^{12} g to the net release of carbon in 1975 (Table 9.3). This was the largest single factor in the net flux.

Regrowth of forests following harvest removes carbon from the atmosphere and stores it in growing vegetation. Using harvest rates derived from the literature and from satellite imagery, the model calculated that regrowing forests accumulated 3.58×10^{12} g C in vegetation in 1975 (Table 9.2). We assumed that a harvested forest returns to its original biomass in 72 years, the time required to support a sustained yield of 1.38 per cent of the standing stock annually.

Disturbance of the forest floor during harvest results in oxidation of soil organic matter over several years (Tamm and Petterson, 1969; Covington,

---

* One of the problems of reconstructing this history is that certain units of measurement are not strictly comparable through time. For example there are not 12 board feet in a cubic foot (cu ft) because the International 1/4 in rule specifies that a board foot is $0.904762 \times (0.22 D^2 - 0.71 D)$, where D is the inside bark diameter at the small end of a four foot length of stem. Thus, the smaller the stem, the fewer the number of board feet per cu ft. For the sawlogs harvested from Maine in 1970 there were only six board feet per cu ft. The difference represents slash, sawdust and other residues which may be used for pulp or other industrial products, may be burned, or may decay over time. Other conversions subject to variability are the number of cubic feet of wood in a 128-cu ft cord (usually 80) and the density of wood itself (FAO, 1979).

Table 9.2   Factors contributing to the net release of carbon from forests of Maine in 1975

Factor	Net flux of carbon in 1975 ($\times 10^{12}$ g)
Oxidation of wood harvested in 1975	0.16
Oxidation of wood products harvested previous to 1975	2.20
Oxidation of logging residues from the 1975 harvests	0.26
Oxidation of logging residues from harvests prior to 1975	4.38
Storage of carbon in forests regrowing after previous harvests	-3.58
Net flux of carbon from soils (includes oxidation and build-up of organic matter)	2.89
Total net release of carbon in 1975 from harvest of forests	6.31

Table 9.3   A comparison of aircraft- and satellite-based estimates of deforestation in Washington

	Aircraft (ha)	Satellite (ha)	Error (%)
	68	68	0
	32	34	6
	50	49	-2
	23	26	13
	6	6	0
	45	36	-20
	40	36	-10
	26	24	-8
	34	29	-15
	8	4	-50
	6	4	-33
	29	35	20
	43	40	-8
	13	10	-23
	15	16	7
	9	9	0
Sum	445	425	4.5

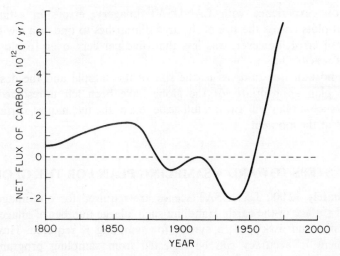

Figure 9.4 Net flux of carbon between the forests of Maine and the atmosphere from 1800 to present. The flux was positive for most of the period but became negative (atmosphere to biota and soils) between 1880 and 1950

1977). The assumption was made for this analysis that 20 per cent of the soil carbon in the top 1.0 m of soil is lost during the 15 years following harvest, and that the carbon content gradually recovers to its original level. Schlesinger's (this volume) estimate of the carbon content of boreal forest soils after 50 years of recovery is 200 MT/ha (20 kg/m^2). The results showed that in 1975 the soils of harvested and regrowing forests were responsible for almost 50 per cent of the computed net flux of carbon from forests (2.89×10^{12} g C) (Table 9.2).

The net annual release of 6.31×10^{12} g C from forests in Maine is a recent phenomenon (Figure 9.4). The increased release of carbon since about 1950 is the result of increased rates of harvests (Ferguson and Kingsley, 1972; Smith, 1972). During the period 1885 to 1950 regrowing forests were responsible for a net accumulation of carbon on land.

The experience, recorded here, in measurement of changes in forests shows part of the process by which remote sensing can be used to provide new, objective data on changes. The results demonstrate how those data can be applied with other data through use of a model to determine fluxes of carbon between the biota and the atmosphere. The combination of techniques developed in Maine and applied later in Washington shows how well satellite imagery can be used routinely to detect and measure changes in the area of forests. Tracts of 400–500 ha can be measured within ± 1–2 per cent; small areas, usually within about 10 per cent. Results of a test of the technique in forested areas of Washington appear in Table 9.3. As would be expected from all previous experience with LANDSAT imagery, errors in estimates of individual plots rise as the size of the area diminishes to one or a few hectares. The overall error, however, was less than five per cent over 16 plots whose total area was 445 ha.

The important questions as to the size of the sample and the design of a sampling plan appropriate for the globe have been left unresolved. While details of such a plan will await a full-scale test of the techniques, certain limits can be set at the moment.

9.5   STEPS TOWARD A SAMPLING PLAN FOR THE GLOBE

Approximately 12 000 LANDSAT scenes are required for one image of the total land surface of the earth. Handling such a large number of images would be expensive and awkward; a system for sampling is required. How much improvement in accuracy can be expected from sampling programmes of various intensities and designs? A detailed, quantitative answer will require experience with the sampling programme adopted, but data and experience available now show that the approach is practical.

Any sampling programme will be stratified to assure that the density of samples is higher in places where change is occurring. A knowledge of the areas of intensive change is needed. There is also a need to define the relationship between the intensity of sampling and the improvement in accuracy. In the following appraisal the ultimate sampling unit is the LANDSAT scene, a large unit (3.4×10^4 km^2 or 3.4×10^6 ha). Scenes will be subsampled in any final programme.

The objective is to use resources most effectively in improving information about rates of change in the area of forests globally. One criterion for the allocation of sampling effort is the uncertainty that exists among current estimates of changes for a region. Sampling might emphasize those regions now marked by uncertainty.

9.5.1   Identifying Critical Regions

Table 9.4 lists the rates of deforestation for 10 regions covering the entire earth. The estimates for tropical and subtropical regions were derived from FAO *Production Yearbooks* (1949–1977) (low estimate) and from Myers (1980) (high estimate). The ranges of carbon fluxes for temperate and boreal regions (non-tropics) were obtained from Armentano and Ralston (1980) and Houghton *et al.* (1983). Restricting the use of LANDSAT to forest reduces the number of scenes from 12 000 to 4200 globally (Table 9.5).

Ninety per cent of the uncertainty in the estimated global carbon flux for 1980 was attributable to five regions (listed in order of least to most certain):

Table 9.4   Range of estimates of deforestation and of carbon flux for the regions of the world

	Deforestation (10⁶ ha/yr)		Flux of carbon (10¹⁵ g/yr)		Number of LANDSAT scenes containing forest
	Low	High	Low	High	
World	27.1	41.5	0	4.5	4260
Tropics	9.3	24.2	1.2	4.1	2128
Non-tropics		17.8	−1.2[b]	0.4	2132
North America[a]		5.7	−0.41[b]	0.003	604
Europe		1.6	−0.075[b]	−0.044	238
USSR[a]		7.8	−0.48[b]	0.5	1040
Pacific developed		0.5	−0.045[b]	0.003	185
China		1.3		0.13	65
Latin America[a]	4.0	7.6	0.73	1.38	847
North America and Middle East	0.26	0.34	0.038	0.056	55
Tropical Africa[a]	2.3	5.7	0.22	0.92	888
South Asia	1.9	2.8	0.17	0.46	114
Southeast Asia[a]	1.6	8.1	0.20	1.47	224

[a] Five regions that account for 90 per cent of the uncertainty in the global flux of carbon.
[b] Armentano and Ralston (1980).

Table 9.5   Number of LANDSAT scenes required for different levels of stratification

	Maximum Number of LANDSAT scenes[a]	
	with 100% sample	with 25% error acceptable
Global areas:		
Total vegetated	12 000	
Total forested	4 200	
Forest areas:		
Five critical regions	3 600	
Total tropical	2 100	360
90 per cent of tropical deforestation	1 200	
Highest rates of tropical deforestation	500	75

[a] Number of scenes calculated by summing, country by country, the scenes containing forest. Because scenes overlap at the edges of countries, some scenes were counted twice and the estimates are high.

Southeast Asia, USSR, Tropical Africa, Latin America, North America (Table 9.4). The forests of these regions are 85 per cent of the world's forests.

It is important to recognize the uncertainties inherent in individual estimates of deforestation such as those based on data from the FAO (1981a,b,c). The recent FAO Tropical Forest Assessment (1981c) estimated deforestation in Burma by two methods. Comparison of aerial photographs from the 1950s with LANDSAT imagery gave an annual estimate of 23 320 ha deforested. An estimate based on numbers of people practising shifting cultivation yielded an estimate of 92 000 ha. The FAO chose the second value as more realistic (FAO, 1981c).

Deforestation is not distributed randomly within regions. Certain countries and areas within countries have particularly high rates of deforestation, and a knowledge of these differences provides the basis for another level of stratification. The FAO Tropical Forest Assessment (1982) gives the rates of deforestation for 76 countries. Countries with rates of deforestation of 15 000 ha per year or higher are listed in Table 9.6. Deforestation in Brazil was divided among the Brazilian states according to data given by Hecht (1982). Thirty-four countries (or states within countries) account for 90 per cent of the FAO estimates of tropical deforestation. The number of LANDSAT scenes required to cover the forests of these countries once is about 1200.

Even within countries there is considerable geographic variation in the rates of deforestation. Myers (personal comm.) claims that the majority of deforestation occurs in zones of intense activity, or fronts (Figure 9.5). About 500 LANDSAT scenes would be required to view these areas. When countries or states are ranked by the rate of deforestation (FAO, 1981a,b,c) per unit area (Table 9.6), the sequence allows one to see the relationship between information and cost. Figure 9.6 shows the number of LANDSAT scenes required to document up to as much as 90 per cent of the total deforestation for the tropics and subtropics (FAO, 1981a,b,c).

9.5.2 The Number of Scenes Required for Accuracy

The discussion of numbers of LANDSAT scenes has been based on the assumption of a 100 per cent sample; that is, a complete analysis of the scenes covering the areas in question. Such a complete analysis would not be carried out; each area would be sampled. The number of samples ordinarily depends on the variance of the population, as shown in the following equation:

$$n = \frac{4\sigma^2}{L^2}$$

where n = the required sample size, σ^2 = the variance of the population (in this case the variance of the area deforested per LANDSAT scene) and L = the

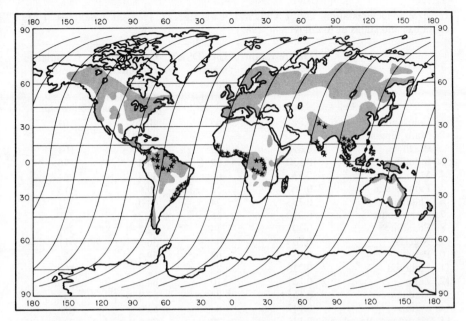

Figure 9.5   Regions of the world containing forests (shaded areas) and regions of intense tropical deforestation (stars). Horizontal lines are latitude: s-shaped lines are the paths (orbits) of LANDSAT in one day. Total coverage of the earth requires 18 days and, hence, 18 times as many orbits. There are 198 LANDSAT scenes in a cell bordered by 0° and 15° latitude and by two of the orbits shown. Data from Myers (personal communication) and the experience of the authors

acceptable error (expressed as a deviation from the mean area deforested per LANDSAT scene).

If an estimate of variance were available, the equation would give the number of scenes required for a 95 per cent probability that the measured rate of deforestation was within the acceptable error (L) of the true rate. To our knowledge an estimate of the variance of deforestation per scene does not exist. We were able to construct a crude estimate, however, by calculating for each of the countries included in the FAO Tropical Forest Assessment (1981a,b,c) the mean rate of deforestation per LANDSAT scene containing forest. The number of forested LANDSAT scenes per country was developed by creating a series of templates of LANDSAT scenes as overlay maps in the *World Atlas of Agriculture* (1969). Thus, for each of the 76 countries an average rate of deforestation per scene was estimated. The variance calculated from these 76 rates is an underestimate because it is based on averages. Nevertheless, we used it to evaluate the equation. We assumed that a 10 per cent error was acceptable; that is, $L = 0.10 \times \text{mean} = 9072$ ha/scene. Thus a sample size of 2250 scenes should give an estimate with 95 per cent probability

Table 9.6 Deforestation and LANDSAT Coverage of Tropical Countries (FAO, 1981a,b,c) and states of Brazil (Hecht, 1982)

	Deforestation (10^6 ha/yr)	LANDSAT scenes containing forest	Cumulative per cent of tropical deforestation	Cumulative number of scenes
1. Colombia	0.80	65	10.7	65
2. Mato Grosso, Brazil	0.608	39	18.9	104
3. Indonesia	0.550	92	26.3	196
4. Mexico	0.53	65	33.4	261
5. Para, Brazil	0.460	52	39.6	313
6. Thailand	0.333	24	44.0	337
7. Ivory Coast	0.310	15	48.2	352
8. Ecuador	0.30	14	52.2	366
9. Nigeria	0.285	45	56.1	411
10. Peru	0.253	68	59.4	479
11. Malaysia	0.230	21	62.5	500
12. Zaire	0.167	110	64.8	610
13. Madagascar	0.165	22	67.0	632
14. India	0.147	86	69.0	718
15. Maranhão, Brazil	0.146	20	70.9	738
16. Laos	0.125	11	72.6	749
17. Venezuela	0.125	46	74.3	795
18. Nicaragua	0.111	9	75.8	804
19. Philippines	0.101	40	77.1	844
20. Rondonia, Brazil	0.099	17	78.5	861
21. Burma	0.0955	31	79.9	892
22. Honduras	0.095	11	81.0	903
23. Nepal	0.084	8	82.1	911
24. Guatemala	0.080	9	83.2	920
25. Cameroon	0.08	31	84.3	951
26. Vietnam	0.065	19	85.2	970
27. Bolivia	0.065	56	86.0	1020
28. Costa Rica	0.060	6	86.8	1032
29. Acre, Brazil	0.043	10	87.4	1042
30. Liberia	0.041	8	88.0	1050
31. Angola	0.04	56	88.5	1106
32. Guinea	0.036	21	89.0	1127
33. Amazonas, Brazil	0.034	69	89.4	1196
34. Panama	0.031	10	89.9	1206
35. Ghana	0.027	20	90.2	1226
36. Sri Lanka	0.025	6	90.6	1232
37. Congo	0.022	13	90.9	1245
38. Papua New Guinea	0.021	26	91.1	1271
39. Kampuchea	0.015	11	91.3	1282

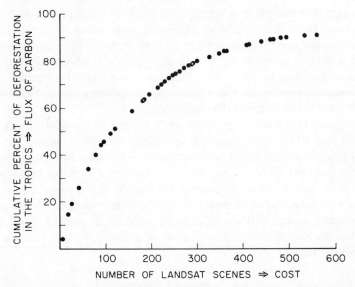

Figure 9.6   Per cent of deforestation observed as a function of LANDSAT scenes analysed. All of the deforestation (rate given by FAO, 1981a,b,c) is assumed to occur in fronts of intense activity

of being within 10 per cent of the true rate. The total number of LANDSAT scenes containing forests in the tropics is less than the required sample size, indicating that the variance is so large as to require use of 100 per cent of the scenes. When the sample size was calculated for the zones of intense deforestation, the required number of scenes was less than that calculated for the 76 tropical countries ($n=475$ scenes), but was, again, of the same magnitude as the total number of scenes for those zones (i.e., about 500) (Table 9.5). Complete coverage appears to be required for an error of 10 per cent or less.

Inasmuch as current estimates of deforestation in the tropics vary by 600 per cent (Houghton *et al.*, 1983), a reduction of the uncertainty to ± 10 per cent may be a greater improvement than is appropriate, at least immediately. If an uncertainty of ± 25 per cent were acceptable, the required sample size for all tropical and subtropical forests and for the zones of intense activity would be 360 and 75 scenes, respectively (Table 9.5). These values are equivalent to a sample of 17 and 15 per cent of the total number of scenes for these levels of stratification. The number of scenes might be reduced still further if subsampling within scenes were to be a part of the overall design. These numbers are small relative to the number of scenes containing forest world-wide. The samples may provide an acceptable estimate of deforestation for the moment, however, because other factors in the flux calculations are of similar uncertainty.

The progressive reduction in the required number of scenes with each step of this analysis (Table 9.5) is due to additional information that provides the basis for stratification. It is important to keep in mind that the information currently available is tentative. Where any sampling scheme is applied, its results must be used to recalculate variance, to change the location of the samples, and to recompute the number of samples required. This iterative process is a continuing part of the sampling plan.

On the basis of what is reported here, sampling of tropical forests might begin with two strata: one, the zones of intense deforestation and the other, the rest of the forest. Information about zones of intense deforestation may be obtainable from the NOAA 7 satellite, for example, which has demonstrated its ability to locate areas of current large-scale disturbance rapidly and inexpensively (Tucker *et al.* in press). Similarly, recent results from the shuttle-based imaging radar (SIR-A) (Elachi *et al.*, 1982) suggest that several approaches may complement one another in a global sampling effort of the type presented here.

9.6 CONCLUSIONS

Satellite imagery can be used effectively to measure current rates of change in the areas of forests globally. The basis of such use, however, must lie in direct measurement of changes, not in indirect analyses based on a comparison of inventories. Such an emphasis enables simplification to the point of avoiding most, but not all, classification of vegetation. The reliance on classification is replaced by detection of change, a much simpler process amenable to routine use of machine techniques.

The use of such data requires a model that incorporates the fundamental principles of ecological succession and the basic data on the carbon content of major types of vegetation. One such model, the MBL-TCM, is used at present. There is always a need for further refinements of the data and assumptions of such models.

Preliminary analyses suggest that a LANDSAT-based appraisal of current changes in the area of forests would be useful if based on a sample of 10 per cent or less of the approximately 12 000 scenes required to cover the land area of the globe.

9.7 REFERENCES

Armentano, T. V., and Ralston, C. W. (1980) The role of temperate zone forests in the global carbon cycle. *Can. J. Forest Res.*, **10**, 53–60.
Brown, S., and Lugo, A. E. (1980) Preliminary estimate of the storage of organic carbon on tropical forest ecosystems. In: Lugo, A. E., Brown, S., and Liegel, B. (eds), *The Role of Tropical Forests on the World Carbon Cycle*, 65–117. Office of Environment, US Department of Energy.
Covington, W. W. (1977) Forest floor organic matter and nutrient content and leaf fall

during secondary succession in northern hardwoods. PhD thesis, Yale University, New Haven. 98 pp.

Covington, W. W. (1981) Changes in forest floor organic matter and nutrient content following clear cutting on northern hardwoods. *Ecology*, **62**, 41–48.

Elachi, C., Brown, W., Cimino, J., Dixon, T., Evans, D., Ford, J., Saunders, R., Breed, C., Masursky, H., McCauley, J., Schaber, G., Dellwig, L., England, A., MacDonald, H., Martin-Kaye, P., and Sabins, F. (1982) Shuttle imaging radar experiment. *Science*, **218**, 996–1003.

FAO Production Yearbook (1949–1977) FAO, Rome, Italy.

FAO (1979) *1977 Yearbook of Forest Products.* FAO, Rome.

FAO (1981a) Tropical forest resource assessment project: Los recursos forestales de la America tropical. FAO, Rome.

FAO (1981b) Tropical forest resources assessment project: Forest resources of Tropical Africa. FAO, Rome.

FAO (1981c) Tropical forest resources assessment project. Forest resources of tropical Asia. FAO, Rome.

FAO (1982) Tropical forest resources. FAO, Rome.

Ferguson, R. H., and Kingsley, N. P. (1972) The Timber Resource of Maine. *U.S.D.A. Forest Service Resource Bull., NE-26.*

Hecht, S. B. (1982) Agroforestry in the Amazon Basin: Practice, theory and limits of a promising land use. In: Hecht, S. B. (ed), *Amazonia: Agriculture and Land Use Research Proceedings of International Conference*, 331–371. CIAT, Cali, Colombia.

Houghton, R. A., Hobbie, J. E., Melillo, J. M., Moore, B., Peterson, B. J., Shaver, G. R., and Woodwell, G. M. (1983) Changes in the carbon content of terrestrial biota and soils between 1860 and 1980: a net release of CO_2 to the atmosphere. *Ecological Monographs*, **53**, 235–262.

Johnson, W. C., and Sharpe, D. M. (1982) Evaluation of merchantable—total biomass conversion ratios used in global carbon budget research. *Canadian Journal of Forest Research* (In press).

Moore, B., Boone, R. D., Hobbie, J. E., Houghton, R. A., Melillo, J. M., Peterson, B. J., Shaver, G. R., Vorosmarty, C. J., and Woodwell, G. M. (1981) A simple model for analysis of the role of terrestrial ecosystems in the global carbon budget. In: B. Bolin (ed), *Carbon Cycle Modelling, SCOPE 16.* John Wiley and Sons, New York.

Myers, N. (1980) *Conversion of Tropical Moist Forests.* National Research Council, Washington, DC.

Persson, R. (1974) World forest resources: review of the world's forest resources in the early 1970's. *Research Notes Nr. 17*, Royal College of Forestry Survey, Stockholm, Sweden.

Riordan, C. J. (1982) Change detection for resource inventories using digital remote sensing data. In: *National Workshop, In: Place Resource Inventories: Principles and Practices*, 278–283, Orono, Maine.

Robinson, J. (1979) A critical review of the change detection and urban classification literature, computer sciences corporation. (SC/TM-79/6235) 51 pp.

Schlesinger, W. H. (1977) Carbon balance in terrestrial detritus. *Ann. Rev. Ecol. Syst.*, **8**, 51–81.

Smith, D. C. (1972) A history of lumbering in Maine 1861–1960. *Univ. of Maine Studies, No. 93.* Univ. of Maine Press, Orono, Maine.

Tamm, C. O., and Pettersson, A. (1969) Studies on nitrogen mobilization in forest soils. *Studia Forestalis Suecia, No. 75.*

Tucker, C. J., Holben, B. N., and Goff, T. E. (1983) Forest clearing in Rondonia, Brazil as detected by NOAA AVHRR eata. NASA GSFC Technical Memorandum 85018, 30 pp.

Whittaker, R. H., and Likens, G. E. (1973) Carbon in the Biota. In: Woodwell, G. M., and Pecan, E. V. (eds), *Carbon and the Biosphere*, 281–302. U.S.A.E.C., Washington, DC.

Wood, R. G. (1971) A History of Lumbering in Maine, 1820–1861. *Maine Studies, No. 33.*

Woodwell, G. M. (1978) The carbon dioxide question. *Scientific American*, **238**, 34–43.

Woodwell, G. M. (1983) Biotic effects on the concentration of atmospheric carbon dioxide: a review and projection. National Academy of Science. (Manuscript).

Woodwell, G. M., Hobbie, J. E., Houghton, R. A., Melillo, J. M., Moore, B., Peterson, B. J., Shaver, G. R., and Stone, T. A. (1983a) Deforestation measured by LANDSAT. Report to the Department of Energy TR005, 62 pp., Washington, DC.

Woodwell, G. M., Hobbie, J. E., Houghton, R. A., Melillo, J. M., Moore, B., Peterson, B. J., and Shaver, G. R. (1983b) The contribution of global deforestation to atmospheric carbon dioxide problem. *Science* (Submitted).

Woodwell, G. M., Whittaker, R. H., Reiners, W. A., Likens, G. E., Delwiche, C. C., and Botkin, D. B. (1978) The biota and the world carbon budget. *Science*, **199**, 141–146.

World Atlas of Agriculture (1969) Instituto Geografico de Agostini, Novara, Italy.

Young, H. E. (1976) Summary and analysis of weight table studies. In: Young, H. E. (ed), *Oslo Biomass Studies*. University of Maine, Orono.

Index

Hueck, mapping vegetation of South
America, 40
Hueck and Siebert, vegetation map of
South America, 30–32, 36, 64
vegetation map of Venezuela, 40
Human roles, Ellenberg's classification
scheme, 41
Humus, 96, 115

IBP, *see* International Biological
Program
Ice structure, effect on return radar
signal, 177
Identification process, remote sensing,
182
India, crop production monitoring, 211
potential vegetation mapping, 42
Indirect gradient analysis, 26
Inceptisols, 94, 95, 97, 98, 99
Industry, role in LACIE, 194
Information tabulation and display, 147
Instrument accuracy, remote sensing, 182
Insect infestations, detection by colour
infra-red aerial photography, 136
Intensity resolution, 162
International Biological Program, 26, 46,
112, 114
International Committee for Vegetation
Mapping, 40

K-band radar, 152
Krajina, British Columbia map, 36,
29–40, 59
Küchler, vegetation map of United
States, 30, 32, 36
vegetation map of Kansas, 32
Küchler's formula, classification of
vegetation architecture, 45–46, 61, 62

L-band radar, 152, 176
LACIE, *see* Large Area Crop Inventory
Experiment
Land cultivation, 119–121
Land use, non-agricultural, 101
Land use changes, 100
LANDSAT data, 149
soil carbon content, 97
see also Forest clearance
Land use types, soil orders, 95
LANDSAT-1, 138
LANDSAT-2, LACIE, 197

LANDSAT-D, 174, 178
thematic mapper, 174–175, 184–185
LANDSAT multispectral scanner system,
139–150, 171, 183, 186, 197, 221–222
capability testing, 191, 195
correlation with radar data, 176
crop production monitoring, 191–217
data analysis, 192, 206–208
limitations, 175, 197, 201
resolution limits, 192, 209, 212–213,
215
sampling plan, 233–238
study of local effects, 187
vegetation classification, 9, 65
vegetation mapping, 65, 122, 138, 161,
222
vegetation stress monitoring, 14, 138,
153, 161, 173, 204, 226–227
see also Multispectral systems; Remote
sensing; Satellite imagery
Landscape classifications, 39, 59
Landscape mapping, potential vegetation,
36–39
Large Area Crop Inventory Experiment,
13–14, 149, 184, 191–217
Layer-diagram method, 47
Life form combinations, 23
Life zone mapping, tropical America, 37
Linear array sensors, 164–165, 178
Litter, 113
Look angle, radar systems, 151
Low temperatures, effect on decom-
position of organic matter, 113

Meteorological data, remote sensing of
crop production, 187–188
Microwave sensing, 175–177
Moisture content of vegetation, radar
signal, 152
Moisture stress, LANDSAT detection,
204
Mollisols, 94, 95, 97, 98
Montane tropical rain forests, mapping,
34
soil organic matter, 113
Mosaic analysis, 26
Mountain soils, 94, 95, 97, 98, 99
MSS, *see* Multispectral scanner systems
Mueller–Bombois classification system,
34, 35, 52
Multiband photography, 133
Multiband linear array sensor, 164, 178